CHANDRAYAAN-3

Dr Ajey Lele is a consultant at Manohar Parrikar Institute for Defence Studies and Analyses (MP-IDSA), New Delhi. His areas of research include issues related to strategic technologies and space security. His published works include *Mission Mars: India's Quest for the Red Planet*, *Asian Space Race: Rhetoric or Reality?* and *Institutions That Shaped Modern India: ISRO*.

CHANDRAYAAN-3
India on the Moon

AJEY LELE

RUPA

Published by
Rupa Publications India Pvt. Ltd 2023
7/16, Ansari Road, Daryaganj
New Delhi 110002

Sales centres:
Bengaluru Chennai
Hyderabad Jaipur Kathmandu
Calcutta Mumbai Prayagraj

Copyright © Ajey Lele 2023
Photos courtesy: ISRO

The views and opinions expressed in this book are the author's own and the facts are as reported by him which have been verified to the extent possible, and the publishers are not in any way liable for the same.

All rights reserved.

No part of this publication may be reproduced, transmitted, or stored in a retrieval system, in any form or by any means, electronic, mechanical, photocopying, recording or otherwise, without the prior permission of the publisher.

P-ISBN: 978-93-5702-686-4

First impression 2023

10 9 8 7 6 5 4 3 2 1

The moral right of the author has been asserted.

Printed in India

This book is sold subject to the condition that it shall not, by way of trade or otherwise, be lent, resold, hired out, or otherwise circulated, without the publisher's prior consent, in any form of binding or cover other than that in which it is published.

Contents

Preface — *vii*

1. India Scripts History — 1
2. The 'Dynamics' of the Moon — 13
3. India's First Two Missions to the Moon — 36
4. Chandrayaan-3: ISRO's Moon Supremacy — 56
5. Learning from Chandrayaan-2's Failure — 76
6. Industry and People — 87
7. Global Moon Agenda — 97
8. Moon Rovers: An Overview — 110
9. India's Moon Tryst — 117

Notes — 120
Index — 131

Preface

India's reputation as a space technology powerhouse took a quantum leap on 23 August 2023, when her Moon mission Chandrayaan-3 made a successful soft landing on the lunar surface. With this, India globally became not only the fourth country after the United States (US), Russia (erstwhile Soviet Union) and China to achieve this feat, but also the first country in the world to successfully land a spacecraft near the south pole of the Moon. This triumph is a result of years of dedication and collaborative efforts of India's scientific community working at the Indian Space Research Organisation (ISRO), an agency which heads India's space programme. This monograph takes an overview of India's Chandrayaan-3 mission and India's overall 'Moon Agenda'.

Chandrayaan-3 is a continuum of India's quest for the moon. To appreciate ISRO's moon pursuit, it is important to recognize the 'processes and threads' associated with this programme. In a broader context, a 'process' is a sequence of steps required to execute any programme, and a 'thread' gets viewed as a subset of a process which runs in the context of the same. It is about a connection between ideas. For appreciating a major idea like a Moon programme, it is important to evaluate the purpose, the science, the technology, the history, the efforts and the challenges. This is not a detailed technical evaluation of the Chandrayaan-3 programme but takes a bird's eye view on India's Moon mission.

As a run-up to Chandrayaan-3, some interesting interviews of subject-matter experts were conducted and made available on the internet. Also, the internet saw a deluge of popular and semi-technical articles before and after this mission. All of these acted as the stimuli for me to undertake this work.

I thank Ambassador Sujan R. Chinoy, Director General, Manohar Parrikar Institute for Defence Studies and Analyses (MP-IDSA), New Delhi for the encouragement to research on various matters related to the domain of space. I am indebted to Yamini Chowdhury from Rupa Publications, India, who pushed me to undertake this work. Last but not the least, my gratitude to my wife Pramada and my son Nipun for always being there.

1

India Scripts History

Since ancient times, people around the world have studied the skies and used their observations and explanations of astronomical phenomena for both religious and practical purposes. There have been some amazing (mostly fictional) stories describing human journeys to the moon and the sun, and about what space means to humans. However, these fictional stories of space travel are hardly fiction anymore. As the twentieth century began, the ancient technology of rockets had advanced to a stage where it became feasible to consider their application for propelling objects to velocities capable of achieving Earth orbit and breaking free from Earth's gravitational pull for deep space travel.

The most exciting year of the twenty-first century, if the marvels in the space domain are considered, is 2022. It is opined that the dawn of a new era in astronomy began this year, when the world got its first look at the full capabilities of the James Webb Space Telescope. The telescope's first full-colour images and spectroscopic data were released on 12 July 2022. These images were of thousands of galaxies, some of which formed nearly 13 to 14 billion years ago.

The year 2023 has also had its own 'space high point'. The Jupiter Icy Moons Explorer (Juice), an interplanetary spacecraft, was launched on 14 April 2023 by the European Space Agency (ESA). The mission aims to study Ganymede, Callisto, and Europa, three of Jupiter's Galilean moons. The spacecraft is expected to reach Jupiter by July 2031. Another mission of interest is NASA's OSIRIS-REx, the first US mission to collect a sample from asteroid Bennu.

It is expected that the analysis of this sample could give some answers to the formations of the planets some 4.5 billion years ago. The OSIRIS-REx spacecraft had collected dust from the asteroid's surface in 2020. The mission returned to Earth on 24 September 2023 with material from the asteroid. Bennu is more than 321 million km away from Earth.[1]

India Scripts History

As of 7 September 2023, three Moon missions have been undertaken by India, Russia (erstwhile Soviet Union) and Japan.

Humble Beginnings

Six decades ago, on 28 August 1963, an estimated 2,50,000 people had gathered at the Lincoln Memorial for the March on Washington for Jobs and Freedom. It was the day when Martin Luther King Jr had spoken about the fight against hatred and bigotry in his famous 'I Have a Dream' speech. India's space programme, a dream of the Indian scientific community, also began during the year 1963. On 21 November 1963, a sounding rocket took off from Thumba, a small village on the outskirts of Trivandrum (now Thiruvananthapuram), Kerala, India. This launch is viewed as the first milestone in modern India's space odyssey.

Thumba was a fishing village with small huts, coconut groves and a tranquil sea. However, it caught the interest of Dr Vikram Sarabhai, the father of India's space programme, for a different reason—a church. In this small hamlet dedicated to St Mary Magdalene, a church was located on Earth's magnetic equator. The magnetic equator is an imaginary line around the globe that joins all the points where a magnetic needle, when freely suspended, is horizontal. It is significant since the magnetic equator is where the equatorial electrojet (EEJ) exists—a stream of electrons whizzing across the sky, about 110–120 km above Earth's surface.

Sounding rockets, which are the initial rockets launched in various space programmes, are employed to conduct research in the domains of physics, astronomy and meteorology by studying electrons. These rockets represent the early stages of development before the establishment of comprehensive space programmes. It was recognized that the geographical coordinates of Thumba at 8°32'34" N and 76°51'32" E offer an optimal setting for investigations focused on the lower atmosphere, upper atmosphere and ionosphere.

At Thumba, the church and the bishop's residence were the only buildings made from cement concrete. This church, and its adjoining premises, were graciously handed over by then bishop Reverend Peter Bernard Pereira and the worshippers of that church to the Indian National Committee for Space Research (INCOSPAR) set up by the Government of India in 1962 (which would later be known as ISRO) to start India's space programme[2]. It could be said that India's space programme started because of the sacrifices made by the people of India. 60 years ago, India started its space programme by firing the US-built Nike-Apache sounding rocket and today, in the year 2023, India has soft landed an indigenously developed rover and lander system on the surface of the Moon.

Availability of high-end technology and financial might to support programmes involving rocket science was necessary to undertake space missions. Obviously, investing in space technology was considered as some distant dream for a developing country like India. It is important to note that since the beginning of its space programme, India had a clarity on why making investments in the technology field which finds 'resonance' only with the developed countries was important. In a speech made in February 1968 at the Thumba Equatorial Rocket Launching Station (later renamed as Vikram Sarabhai Space Centre), Vikram Sarabhai said: 'There are some who question the relevance of space activities in a developing nation. To us, there is no ambiguity of purpose. We do not have the fantasy of competing with the economically advanced nations in the explorations of the Moon or the planets or manned space flight. But we are convinced that if we are to play a meaningful role nationally, and in the comity of nations, we must be second to none in the application of advanced technologies to the real problems of man and society[3].'

Dr Sarabhai's Vision

For the last 60 years, India has respected the vision of Dr Sarabhai. India's primary focus on space programmes has been on using satellite technology for the purpose of socio-economic development. Unfortunately, Dr Sarabhai died on 30 December 1971, at the age of 52. Subsequently, it was Professor Satish Dhawan (chairman of ISRO during 1972–1984), who gave the foundation to India's space dream. It was he who laid out ISRO's long term programmes of applications and science, and the technologies needed to execute those programmes self-reliantly. It could be inferred from various speeches of Dr Sarabhai, that he believed programmes like Moon missions were glamorous programmes meant for developed nations. Author and retired ISRO scientist Dr V Siddhartha (also worked in Defence Research and Development Organisation) had mentioned that Professor Dhawan was also sceptical about the scientific, or even technological value, of manned missions. However, it is important to understand that both Dr Sarabhai and Professor Dhawan were at the helm of affairs during the period of Cold War when then superpowers, namely the US and the Soviet Union, were using the space domain as a tool for geopolitics. The Soviet cosmonaut Yuri Gagarin becoming the first human to orbit the Earth on 12 April 1961 and US astronaut Neil Armstrong becoming the first human to walk on the Moon on 20 July 1969 definitely had something to do with science and technology, but there was a major geopolitical messaging associated with this. During the end of the twentieth century, the scientific relevance of undertaking missions to the Moon and other planets started becoming clear, and India did a mid-course correction with their space policies.

India became a spacefaring nation in 1980 when it demonstrated its capability to launch a satellite from its soil by using an Indian rocket launch system. Over the years, ISRO has developed purpose-specific launch vehicles. India has two important vehicles for undertaking satellite launching—Polar Satellite Launch Vehicle (PSLV) and Geosynchronous Satellite Launch Vehicle (GSLV). Various variants of these launch vehicles are available to undertake launching of satellites in various orbits (from low Earth orbit to geostationary orbit). In addition, ISRO has recently developed a Small Satellite Launch Vehicle (SSLV) for launching mini, micro and nano satellites. Since the beginning, Indian space programme's focus has been to launch satellites mainly for remote sensing purposes. Here, the aim is to get data on natural resources and use it for the purposes of land, water and forestry management. India, being an agricultural economy, has also focused on development and launching of satellites for meteorological purposes. India has its own regional space-based navigation system.

It is said that Dr Krishnaswamy Kasturirangan, who was the ISRO chairman for nine years (1994-2003), thought that ISRO could play a small role in India's ambition to become a superpower. The idea of a Moon orbiter was floating during his tenure. Going to the Moon was not a utopian idea to start with. Some initial assessments of ISRO's capabilities were carried out. At the time, ISRO knew how to design satellites for geostationary orbits which could carry plenty of fuel on board[4]. By 2001–02, PSLV had already undertaken more than five successful launches. So, the basic infrastructure seemed to be in place and the only change required was adjusting a geostationary satellite for the Moon. Technical assessment showed that the PSLV rocket could provide an Earth-bound orbit beyond which the fuel on the spacecraft could be used to go to the Moon and perform orbital capture[5].

India's Moon Missions

In 2006, a meeting of top scientists from the country was held at Bangalore (ISRO's headquarter is located here) to discuss the future of India's space programme. After this meeting, it was observed that ISRO started taking a major interest towards conducting missions in deep space. Much before this meeting, the idea of an Indian scientific mission to the Moon was first raised in 1999 at a meeting of the Indian Academy of Sciences. In 2000, this idea was also pushed by the Astronautical Society of India (ASI)[6]. Since the 1980s to the 2000s, ISRO had established itself as an agency capable of undertaking development and launching operational spacecraft systems for survey and management of natural resources, meteorological services and satellite communication. There was a realization that if the technologies, developed and available at ISRO, were used and modified correctly, ISRO could embark on planetary missions with meticulously planned scientific objectives. PSLV's capability was judged from the point of view of undertaking Moon missions and other terrestrial planets. It was realized that this vehicle could be used for launching a Moon mission to a certain altitude and subsequently additional innovative ideas (techniques for raising the altitude of the craft) could be put in place. Finally, based on the recommendations of the scientific community and as a first major initiative, a National Lunar Mission Task Force was constituted by ISRO. It comprised leading scientists and technologists from all over the country for considering and assessing the possible configuration and feasibility of taking up an Indian Moon Mission.

The study report of the task team was reviewed in April 2003 by a peer group of about 100 eminent scientists from various relevant fields of planetary and space physics, earth sciences, geology, astronomy and cosmology. After detailed discussions, the participants unanimously recommended that India should undertake the Moon mission. The recommendations are summarized as follows:

- The Indian Moon mission assumes significance in the context of the international scientific community considering several exciting missions in planetary exploration in the new millennium.
- ISRO has the necessary expertise to develop and launch the Moon mission with imaginative features and it would be different from the past missions. Hence ISRO should go ahead with the project approval and implementation.

- Apart from technological and scientific gains, it would provide the needed thrust to basic science and engineering research in the country. The project would help the return of young talents to the arena of fundamental research.
- The academia, in particular the university scientists, would find participation in such a project intellectually rewarding. In this context, the scientific objectives would need further refinement to include other innovative ideas from a broader scientific community through announcement of opportunity, etc.

Thereupon, the Government of India approved ISRO's proposal for the first Indian Moon mission called Chandrayaan-1.

Chandrayaan, a word from the ancient Indian language Sanskrit, is a combination of two words: *Chandra* and *Yaan*. In Sanskrit, Chandra means moon and Yaan means craft or a vehicle. Hence, India's Moon mission is called 'Moon Craft' meaning Chandrayaan. The first Moon mission was operated between October 2008 and August 2009. The second one, which was a partial success, was launched during July 2019, and the orbiter part of this mission is still working as of 15 September 2023 and is expected to function till around 2026. And now, the third Moon mission has successfully landed on the lunar surface and its rover and lander functioned as designed, for one lunar day which is equivalent to 14 Earth days.

2

The 'Dynamics' of the Moon

About the Moon

In Latin, the moon is called *Luna*, hence the word lunar gets used as a main adjective for all things related to the Moon. It is Earth's only natural satellite and is the only place beyond earth, where humans have set foot, so far. One obvious question regarding the moon is 'When did we first feel the presence of the moon?' There is no perfect answer to this question. It could be said that the presence of the moon must have been acknowledged by a form of life on earth having an ability to see that much of a distance. This sighting of the moon could have happened billions of years ago.

In ancient China, the Moon's motion was judiciously chronicled as part of a grand structure of astrological thought. In both China and West Asia, some accurate observations helped in the prediction of eclipses, and this data was of great assistance in later years to the scientists interested in tracing the history of the Earth-Moon system. It was Greek philosopher Anaxagoras around 400 BC, who reasoned that the Sun and the Moon were both spherical rocks. He also realized that we can see the Moon in the first place as it does not emit its own light but reflects the light of the Sun.

The Moon rotates at the same rate that it revolves around Earth. Hence, the same hemisphere faces Earth all the time. The Moon makes a complete orbit around Earth in 27 Earth days. Because Earth is moving as well as rotating on its axis as it orbits the Sun, from our perspective, the Moon appears to orbit us every 29 days. The Greek astronomer and mathematician Hipparchus observed Earth's round shadow creeping across the Moon during a lunar eclipse and concluded that Earth must be spherical and that the moon was an independent world. He also correctly described the Moon's phases and accurately estimated the distance between the two bodies.

The Moon has eight phases in a lunar month.

The 'Dynamics' of the Moon

Half of the Moon's surface is continually illuminated by sunlight. As the Moon orbits Earth, the extent of its lit-up side which can be seen from the Earth keeps changing. The eight Moon phases are divided into four primary and four intermediate (waxing and waning) phases. They are:

1. New Moon
2. Waxing Crescent Moon
3. First Quarter Moon
4. Waxing Gibbous Moon
5. Full Moon
6. Waning Gibbous Moon
7. Third Quarter Moon
8. Waning Crescent Moon

The Moon has a radius of about 1,737.5 km and its distance from the Earth is about 3,84,400 km. Roughly, 30 Earth-sized planets could fit in between Earth and the Moon. The Moon's total width is analogous to that of the country of Australia. The surface area of the Moon is roughly 38 million square km and its weight is estimated to be 1/81 of Earth's. Historically, when the Moon was just formed, it was in a much closer orbit to Earth. This means, it would have appeared much larger in the sky then. Over a period, the Moon's orbit has expanded further outward increasing the length of the orbit at the same time. It has been observed that the Moon is slowly moving away from Earth; say about an inch farther away each year.

The surface of the Moon witnesses extreme temperatures. During daytime on the Moon (when sunlight is there), the temperature reaches to about 130°C, and in darkness it becomes as low as –200°C. Such intense heat and cold are common on the Moon. Here, the variations in temperature are extreme because the Moon doesn't have its own atmosphere, unlike Earth.

Since humans have been able to visit the Moon's surface and collect rock and sand samples, it is possible to know more about the origin of the Moon. These samples have been tested by using isotope dating. Based on such testing, it has been inferred that the Moon formed around 50 million years after the proposed origins of the Solar System itself. Possibly, it got formed around 4.51 billion years ago, sometime after the formation of Earth. Although there are some interesting theories regarding the origin and evolution of the Moon, they have already been negated by the scientists.

In the early 1600s, the German astronomer Johannes Kepler, based on his observations, formulated the laws governing planetary motion. In 1609–10,

The 'Dynamics' of the Moon

Galileo began his telescopic observations that totally changed the human understanding of the Moon. For more than three centuries—seventeenth, eighteenth and nineteenth—researchers examined different theories on lunar origin and attempted to match them with the actual observations.

There are three lunar origin theories: Coaccretion, Fission and Capture. According to the Coaccretion theory, a primordial cloud of gas and dust gave birth to the Moon and Earth together. This proposition, however, fails to explain the large angular momentum of the present system. In Fission theory, a fluid proto-Earth (early Earth, perhaps in its first one billion years of existence) began rotating so rapidly that it flung off a mass of material that formed the Moon. Though convincing, the theory eventually failed when examined in detail; scientists were unable to find a combination of properties for a spinning proto-Earth that would eject the right kind of proto-Moon.

According to Capture theory, the Moon formed in a different place in the Solar System and was later trapped by the strong gravitational field of Earth. This scenario had many takers for a long time but the circumstances desirable in celestial mechanics to break a passing Moon into just the right orbit always looked unlikely.

By mid-twentieth century, scientists started

thinking of an additional dimension for a viable lunar origin theory. They started looking at the geological and chemical evidence too. It was thought that the Moon is much less dense than Earth since it contains significantly less iron. This large chemical difference argued contrary to a common origin for the two bodies.

On the other hand, various independent origin theories had their own complications. Even the human mission undertaken to the Moon (Apollo, 1969–72) did not provide any clue. Finally, around the early 1980s, a model of 'the giant-impact hypothesis' emerged that eventually gained the support of most lunar scientists. As per this, the proto-Earth, shortly after its formation from the Solar Nebula about 4.6 billion years ago, was struck by a glancing blow of another big body[7]. Prior to the impact, both bodies previously had undergone differentiation into core and mantle. The titanic collision ejected a cloud of fragments. This aggregated into an almost full ring around Earth and then split-up into a proto-Moon. The ejected matter consisted mainly of mantle material from the colliding body and the proto-Earth, and it experienced enormous heating from the collision. Hence, the resulting proto-Moon was highly washed-out in volatiles and moderately depleted in iron.

The 'Dynamics' of the Moon

The proto-Moon in the debris cloud, could have quickly swept up the remaining fragments. Then, over a period of 100 million years or so, some other changes in the body of the proto-Moon could have taken place (owing to collisions, asteroid hits, volcanic events, lava outflows etc.). Subsequently, the Moon must have cooled and solidified.

To unravel the history of this period, scientists have applied modern analytic techniques to lunar rock samples. The mare basalts show a wide range of chemical and mineral compositions reflecting different conditions in the deep regions of the mantle where, presumably because of heating from radioactive elements in the rock, primordial lunar materials were partly re-melted and fractionated so that the lavas carried unique trace-element signatures up to the surface. By studying the past events and processes reflected in the mineral, chemical and isotopic properties of these rocks, lunar scientists have slowly built a picture of a variegated Moon. Their findings have provided valuable background information for Earth and spacecraft-based efforts to map how the content of important materials varies over the lunar surface. During the last few years, scientists have been able to reach such conclusions based on computer modelling of the collision and other events.

The geology of the Moon is much different from that of Earth. There is very little atmosphere over the Moon causing of the landmass due to weather. It has lower gravity and does not possess any form of plate tectonics. The complex geomorphology of the lunar surface got formed owing to a combination of processes like impact cratering and volcanism. The Moon has undergone planetary differentiation, therefore containing a crust, mantle and core. The Soviet Union undertook a robotic equipment landing mission on the lunar surface during 1959.

Over the last six decades or so, few manned missions and various unmanned missions have reached the lunar surface. Today, there are samples available with a known geological context for analysis. Various Apollo missions of NASA brought back 381.7 kg of lunar surface material and the unmanned Soviet Luna programme returned with 0.3 kg of lunar material[8]. More recently, China's Chang'e-5 lunar sample return mission (2020) brought back 1.73 kg of lunar material. Also, there have been various orbiting lunar spacecraft and flyby missions that have collected some amount of information about the nature of the lunar surface and the stuff beneath it. In addition, for centuries, various telescopic observations have been taken. These have helped to generate a large volume

The 'Dynamics' of the Moon

of data to undertake geological studies. However, this does not mean that we have all the geological information available.

There are six major epochs in the geological history of the Moon. It is called the lunar geologic time scale. Starting about 4.5 billion years ago, the newly formed Moon was in a molten state. This early period in the Moon's life was marked by the crystallization of the near global magma ocean, possibly at a depth of about 500 km or greater. The first minerals to have formed in this ocean were the iron and magnesium silicates olivine and pyroxene. A vast majority of this ocean crystallized in the next 100 million years or even less. Quickly, after the lunar crust formed, different types of magmas gave rise to the formation of Mg-suite norites and troctolites (types of rocks). The first rocks brought back by Apollo 11 were basalts and their analysis led to this bold hypothesis.

There has been a detailed assessment in regards to lunar rocks and soil, and scientists have catalogued various region-specific information. There is a clarity about the chemical and mineral properties of lunar rocks and soils. The minerals and materials on lunar surface and below are of igneous origin, but their melting and crystallization history is complex. The materials formed of various minerals are classified into four main groups:

1. Basaltic volcanics, the rocks forming the maria[9]
2. Pristine highland rocks uncontaminated by impact mixing
3. Breccias and impact melts, formed by impacts that disassembled and reassembled mixtures of rocks
4. Soils, defined as unconsolidated aggregates of particles less than 1 cm in size, derived from all the rock types

Based on all available information, scientists have created a picture of the Moon as a layered body comprising a low-density crust (60 to 100 km in thickness), overlying a denser mantle, which constitutes the great majority of the Moon's volume. At the centre, there probably is a small iron-rich metallic core with a radius of about 350 km[10]. However, there are still various uncertainties about the exact knowledge of the lunar interior and also about its exact origin and evolution.

On the lunar exterior, a thick layer of regolith covering the entire lunar surface is most visible. It contains all dimensions of material, from big boulders to sub-micron dust particles. At times, the word regolith gets used interchangeably with lunar soil.

Regolith is a fragmental and unconsolidated rock material. It is being hit continuously by large as well as small meteoroids and the constant shelling of charged particles from the Sun and stars. The regolith is about 4–5 m thick in mare regions and 10–15 m in highland areas. Below the regolith is a region of large blocks of material, large-scale ejecta and brecciated bedrock, known as the megaregolith. Lunar soil makes up a subcentimetre or submillimetre fraction of the regolith. Lunar dust refers to only the finest fractions of the soil, less than 10 or 20 microns. Lunar soil is much different from the earth's soil. It contains no organic matter and is not formed through biological or chemical means. Instead, its formation is a result of the mechanical comminution from meteoroids and interaction with the solar wind and other energetic particles. This soil has no exposure to wind and water, hence it is not found in any organized form and its grains are sharp with fresh fractured surfaces[11].

Why Moon?

Before debating about the reasons for humans to go back to the Moon, it is important to recognize those who have already been there. In 1959, the Soviet

Union's Luna 1 was the first spacecraft to reach the Moon. However, a decade later, a major revolution in the moon affairs happened. History was made in 1969 with the US astronaut Neil Armstrong becoming the first human to walk on the lunar surface. This success by NASA remains unparalleled in the field of space sciences and technologies. Till date, only 12 individuals have walked on the surface of the Moon:

1. Neil Armstrong: Apollo 11 (1969)
2. Edwin 'Buzz' Aldrin: Apollo 11 (1969)
3. Charles 'Pete' Conrad: Apollo 12 (1969)
4. Alan Bean: Apollo 12 (1969)
5. Alan B. Shepard Jr: Apollo 14 (1971)
6. Edgar D. Mitchell: Apollo 14 (1971)
7. David R. Scott: Apollo 15 (1971)
8. James B. Irwin: Apollo 15 (1971)
9. John W. Young: Apollo 16 (1972)
10. Charles M. Duke: Apollo 16 (1972)
11. Eugene Cernan: Apollo 17 (1972)
12. Harrison H. Schmitt: Apollo 17 (1972)

On 20 July 1969, a global audience of some 600 million watched or listened to the progress of Apollo 11 mission with shock and awe! It was a great leap for mankind, and for science, but there was a major political significance to it. Neil Armstrong visiting the

The 'Dynamics' of the Moon

Moon was a geopolitical statement made by the US to demonstrate their technological superiority. The 1960s was a period when possibly the second peak of the Cold War was visible. Earlier, the Cold War had reached its peak during 1948–53. In this period, the Soviets were unsuccessful in blocking the Western-held sectors of West Berlin. However, they had successfully demonstrated their nuclear capabilities by undertaking the first nuclear test during 1949 and, in the same year, the Chinese communists came to power. The Korean War was at its peak during the same period. This period also witnessed the formation of the North Atlantic Treaty Organization (NATO) on 4 April 1949. This military alliance between US and Europe is a major quandary for Russia even today.

During 1961, the US had supported an unsuccessful invasion of Cuba by Cubans exiled in the US. Both the superpowers, US and Soviet Russia, were developing intercontinental ballistic missiles around the same period and 1962 witnessed the Cuban missile crisis which almost brought these two power blocks on the brink of a war. While on the space front during 1961, the Soviets were successful in sending the first man, Yuri Gagarin, to outer space, challenging US's technological superiority. This had taken the Cold War stand-off beyond sea, land and air domains. Space accomplishments were no longer only about national prestige and Neil Armstrong reaching the Moon was a defining part of the push for US global supremacy.

The final human mission to the Moon happened during 1972 and since then, no humans have left the orbit of the Earth. On their part, the Soviets did try to imitate the US in the field of human Moon missions. They had never announced their ambitions officially. However, it is known that they were unable to develop a rocket required for such missions. It would be incorrect to bracket the entire Apollo programme as only a political programme. Some science was definitely on the NASA agenda.

This programme played an important role in getting the lunar rock and soil samples back to Earth. There has been much learning from this programme which is actually helping the US and others towards planning their twenty-first century Moon programme. However, during the 1970s, the US government was clear about their overall expectations from this programme. After accomplishing the geopolitical goal, the Apollo programme was discontinued post 1972 for two reasons: one, it was a costly programme and two, there was no major scientific agenda in place for continuing the programme. Post 1972, the Moon was put in a cold storage. The Soviets also abandoned

The 'Dynamics' of the Moon

their Luna program in 1976. However, this is not to say that nothing happened towards absorbing the unknown secrets of the Moon. NASA and the Japanese agency Institute of Space and Astronautical Science (ISAS), either jointly or individually, did undertake flyby and orbiter missions till the end of twentieth century.

With the beginning of the twenty-first century, fresh interest was(and is) found brewing towards undertaking missions to the Moon. Now, beyond US and the Russians, new space players have arrived. Here, it is important not to overlook the overall developments, which have happened in the domain of space since the 1960s and 70s period. The world has seen rapid technological growth in various fields. China has emerged as one of the most competent space players challenging the US space hegemony. Beyond the European Space Agency (ESA), individual countries like Japan, India and Israel are making their presence felt. Countries like South Korea, Iran, UAE, Saudi Arabia, Australia, Canada and others are increasing their investments in the space domain and opting for innovative programmes. Some of them have clearly identified their interests towards undertaking missions to the Moon.

The world is making a shift from automatic systems

to autonomous systems. Different technological fields from communications to photography have witnessed technology disruption. Technologies in the domain of space, from rocket launchers to sensors, have evolved. There is an increasing awareness regarding the need for the global society to invest more towards scientific research, development and innovations. Ideas like investments in STEM (science, technology, engineering and mathematics) as an interdisciplinary approach towards learning are panning out and helping space sciences and technologies to grow rapidly. Over the years, the world is becoming increasingly dependent on space technologies for the conduct of various routine activities. Many countries have started making more investments in the space domain. Major private agencies also have their own Moon programmes and some are even trying to undertake human Moon missions.

It would be incorrect to conclude that in the twenty-first century, the geopolitical attraction for the Moon has totally withered away. States have understood that blind investment in any field of technology just for prestige or one-upmanship is suicidal. Today, States understand that financial investments in the space domain should not be done solely with a geopolitical purpose and there

The 'Dynamics' of the Moon

should be some scientific and economic advantages behind such investments. All this has led the States to recalibrate their moon agendas. It is known that projects like Moon missions have long gestation periods. Nobody is going to the Moon for some immediate tactical advantages (if there are any, then it is a bonus). It is implicit that such projects are strategic in nature.

There is also another practical aspect that has renewed the world's interest in the moon. Earth's natural resources include air, water, soil, minerals, fuels, plants and animals. Owing to various reasons including rise in global population, disparities among the States from Global North and Global South and lopsided consumption of resources by different States, it is evident that the mineral resources have been managed carelessly for all these years. The world has also realized that conservation alone will not be enough to sustain the future generations. Additionally, the challenges owing to climate change demand immediate remedial measures. Under these circumstances, we need to look for resources somewhere else and the Moon becomes an interesting option in this regard. It may take decades for humans to get enough resources from the Moon to Earth. What we are witnessing today is the beginning of that pursuit. In a sense, it could be said that the modern-day 'moon race' is driven primarily by the economics of moon mining.

It is important to realize that human interest in mining the Moon has two purposes: one, to get the minerals back to Earth for satisfying the needs of the day and near future; two, to use the minerals and other material available on the Moon for the purposes of in-situ use. A few States have ambitions of establishing human colonies over the Moon and such material would come handy in that context. Any extended human residence on the Moon is thinkable if missions would use the local resources. Today, the Moon gets viewed as a staging-ground for humans to undertake missions to Mars and beyond. Obviously, to establish an active base on the Moon for such purposes would require the establishment of massive infrastructure and it would not be viable to carry the hardware required for this purpose from Earth. Some common minerals, including basalt, iron, quartz and silicon are known to be present on the Moon and they could be used for building shelters, solar panels and other required infrastructure. Transporting oxygen from Earth would not be cost-effective in case of long-term human sustenance on the Moon. Hence, processes need to be evolved for in-situ generation

The 'Dynamics' of the Moon

of oxygen. All this is not going to be an easy task.

Finding the required material is only the first step. A major effort would be required to devise and implement innovative mining industry practices on the Moon. In addition, mineral processing units would be required to be built on the lunar surface for specific in-situ usage and for building structures. Presently, the focus is towards development of 3D printing (additive manufacturing) techniques for developing infrastructure on the lunar surface. The present-day Moon missions are trying to get information on the possibilities of Moonquakes (similar to Earthquakes) and meteorite strikes on the Moon. Also, information is required about the possibilities of radiation bursts and electromagnetic interferences in order to plan the development of infrastructure. Further, secured and robust commutation facilities are mandatory both for robotic and human missions.

In recent times, the presence of Helium-3[12] on the lunar surface has been the biggest attraction for planning the Moon missions. Presence of Helium-3 on Earth's surface is rare. It is available in abundance on the lunar surface and could hypothetically be cheaper to mine from the Moon. This substance is an attractive fuel for future nuclear fusion reactors. It can replace the existing uranium/thorium route for energy generation. It is said that if we succeed to get a Boeing-737 load of Helium-3 back on Earth, then it could help to resolve the energy crisis for the entire world for a decade provided a commercial Helium-3 reactor is ever built. As per some estimates, a kg of Helium-3 could fetch $3,000,000. One of the goals of near-future Moon missions will probably be to map Helium-3 deposits for future mining and exploitation. The next step would be to undertake mining on the Moon and have a facility built (rocket systems and spacecraft) to take the material back to Earth.

Presently, China has a virtual monopoly on Earth's dwindling supplies of the Rare Earth Elements (RREs) and Rare Earth Metals (REMs). The Moon is known to have a good presence of these elements. Requirement of these metals for modern electronics, batteries, renewables and electric motors is well-known. They would be required both for in-situ use and also for the use of Earth.

Other scarce materials have also been discovered. On the Moon, there are precious metals such as platinum, palladium and rhodium which are highly conductive and could be used in electronics. As per scientific assessment, the titanium ore discovered on the Moon is 10 times richer than that available on Earth. This ore, when mixed with aluminium or iron would make a useful alloy that is lightweight, corrosion-resistant, incredibly strong and resistant to extreme temperatures. It could be used to build a variety of items such as engines, medical implants and structural frames.

One natural resource uniquely available on the Moon is its polar environment (south pole) and it is expected that major deposits of water (not in liquid form) would be available over there. This is because the Moon's axis is nearly perpendicular to the plane of the ecliptic—therefore, sunlight is always horizontal at the lunar poles. There are some areas close to the south pole where no sunlight ever reaches. Hence, it is expected that in certain areas like crater bottoms and underneath the rocks, where the temperature is around -200 to $-233°C$, some deposits of water would be available. For continued human survival on the Moon, local availability of water is a must. It is expected that the Moon's southern polar region could have an estimated 600 billion kg of water ice.

The future of planetary spaceflights or missions in deep space would have major dependence on water available on the Moon. It would be economical (less fuel would be required) for making the spacecraft to fly to other planets from the surface of the Moon. Also, in future, spacecraft would be flying back to

Earth from the Moon with the 'mineral load'. It needs to be noted that Earth's atmosphere and gravitational pull demand the use of tons of fuel per second during a rocket launch. The Moon has no atmosphere and less gravity. Hence, undertaking launches from the lunar surface would be much more cost-effective. But, for all this, fuel should be available on the Moon, and it should be produced in-situ. When we split water into hydrogen and oxygen, and then liquefy those constituents, we can generate rocket fuel. However, all these are theoretical calculations, and much work would be required to materialize these ideas. It needs to be kept in mind while developing the technologies that there could only be a finite quantity of water available on the Moon. Water should be treated as a recyclable resource for life support (drinking water and breathable oxygen). Today, on the International Space Station (ISS), astronauts' urine gets treated and processed for converting it into drinkable water.

From a scientific perspective, the lunar poles could be the best sites for certain astronomical observations. To observe objects in the cosmos that radiate in the infrared and millimetre-wavelength regions of the spectrum, astronomers need telescopes and detectors that are cold enough to limit the interference generated by the instrument's own heat. Presently, such telescopes launched into space are known to carry cryogenic coolants, which eventually run out. Hence, if a telescope is permanently sited in a lunar polar cold region and insulated from local heat sources, then we could get continuous observations from around half the sky. Another telescope on the other pole could help us cover the entire sky, however, it needs to be noted that both the regions would have different temperature profiles.

3

India's First Two Missions to the Moon

Deep Space Missions

It is difficult to answer what space is. Possibly, the word comes from the Latin term *spatium*, meaning 'space' in English. The word can be broadly defined as the emptiness between two things—it is possible to put something in it because nothing is currently there. We can find space anywhere: on a table, on paper, in a cupboard or on a bus. However, space, as in 'outer space', the abode of the stars, is supposed to be largely empty. Hence, the definition of outer space is commonly perceived synonymous with the literal definition of the word 'space' although the actual definition is quite complex. If anyone is asked where outer space is, one would point upwards. However, it might need some rethinking as terms like air space complicate the definition of outer space even more. American science journalist Nadia Drake writes, 'No one really knows where "air space" ends and "outer space" begins [...] International treaties define "space" as being free for exploration and use by all, but the same is not true of the sovereign air space above nations. The laws governing air space and outer space are different; flying a satellite 55 miles above China is just fine if space begins at 50 miles up, but define the edge at 60 miles, and you might find your satellite being treated as an act of military aggression.' She broadly defines outer space as 'the point where orbital dynamic forces become more important than aerodynamic forces, or where the atmosphere alone is not enough to support a flying vessel at suborbital speeds.'[13]

This point of division is famously known as the Kármán line, the altitude between earth's atmosphere and outer space at which Earth's atmosphere ends and outer space begins. Further, space is a vacuum where mostly the molecules are not close enough to transmit sound between them. However, the space is not empty and there are gases and dust and other bits of matter floating around 'emptier' areas of the universe, while more crowded regions can host planets, stars and galaxies. There is no the exact estimate about how big space is. Over the years, humans have identified various galaxies and are aware that it is impossible to reach them with the present level of technology available. We measure long distances in space in 'light-years,' representing the distance it takes for light to travel in a year (roughly 5.8 trillion miles, or 9.3 trillion km)[14]. It is important to note that not only these distances, but other relatively shorter distances are also difficult to reach. Humans are finding it difficult to reach the planets which belong to our own Solar System. The challenge is not only the distance, but also other atmospheric challenges faced during the travel and the severe weather and terrain conditions (including the lack of oxygen and water) on the planet.

For centuries, humans were only depending on the science of astronomy to understand about the stars and planets. The invention of the telescope added much to the knowledge of early astronomy. The concept of deep space could be said to have its origins in early astronomy. The definition of deep space seems contextual. For some, it refers to interstellar space, while others consider anything beyond the Earth's atmosphere deep space. Also, some use the term to reference anything beyond the Solar System. One formal definition provided by the International Telecommunication Union (ITU) describes 'deep space' starting at a distance of 2 million km from the Earth's surface while NASA Deep Space Network has described deep space starting at a distance of 16,000 to 32,000 km from Earth. So, as per ITU's definition, the Moon does not qualify to be in deep space (but Mars does)[15]. However, for some, the Moon is deep space. Hence, beyond the definitions, it is not entirely wrong to say that any satellite (or a probe) which travels to a distance of 1,00,000 km or more from the Earth's surface is known to have entered the region of deep space. Without delving into any so-called definitional inadequacies, for the purpose of this work, India going to the Moon is being viewed as India's foray into deep space.

The minimum distance from Earth to Mars is about 54.6 million km. The farthest apart they can be is about 401 million km. The average distance is about 225 million km. The average distance between Earth and Mars is over 200 times as far as it is from the Earth to the Moon. The longest distance a satellite gets placed usually is the geostationary orbit, at an altitude of around 36,000 km above the Earth's surface. Even today, some of the spacefaring countries do not have the capability to put satellites into the geostationary orbit. Few countries, including India, have been able to make their way towards deep space. ISRO has successfully undertaken missions to the Moon and Mars. This chapter covers India's Chandrayaan-1 and Chandrayaan-2 missions.

Chandrayaan-1

Chandrayaan-1 was India's first lunar mission and also India's first deep space mission. It operated for around nine months (between October 2008 and August 2009, a total of 312 days) against the projected lifespan of two years. However, as per ISRO, it did carry out most of its planned tasks (amounting to almost 95 per cent) during these nine months. The estimated cost for the project was ₹3,86,00,00,000. One of the major achievements of ISRO during this mission was the discovery of water molecules on the Moon's surface. The mission included a lunar orbiter and an impactor. The lunar orbiter was a satellite which was put in the vicinity of the moon and it took pictures for nine months. It had various sensors on-board for undertaking specific assessments about the Moon's surface.

India's First Two Missions to the Moon

The Chandrayaan-1 mission was aimed at high-resolution remote sensing of the Moon in visible, near infrared, low energy X-ray and high-energy X-ray regions. Specifically, the objectives were:

- To prepare a three-dimensional atlas (with a high spatial and altitude resolution of 5–10 m) of both near and far side of the Moon.
- To conduct chemical and mineralogical mapping of the entire lunar surface for distribution of mineral and chemical elements such as magnesium, aluminium, silicon, calcium, iron and titanium as well as high atomic number elements such as radon, uranium and thorium with high spatial resolution.

The simultaneous photo-geological, mineralogical and chemical mapping through Chandrayaan-1 mission was expected to enable identification of different geological units to infer the early evolutionary history of the Moon. The chemical mapping was to determine the stratigraphy and nature of the Moon's crust and thereby test certain aspects of magma ocean hypothesis. This in turn was expected to help determine the compositions of impactors that bombarded the Moon during its early evolution, which is also relevant to the formation of the Earth[16].

Chandrayaan-1 spacecraft was about the size of a refrigerator, with a dry weight of about 525 kg and was powered by a solar array that charged lithium-ion batteries on board. It was launched on 22 October 2008 from the Indian soil aboard a PSLV rocket. It was inserted into lunar orbit on 8 November 2002. The spacecraft released its Moon Impact Probe (MIP) on 14 November, which crashed into the Moon on the same day.

Chandrayaan-1 completed 3,400 orbits around the Moon and continued transmitting data until 23 August 2009, after which communication with the orbiter was permanently lost. It is noteworthy that Chandrayaan-1 was dispatched to the Moon through a sequence of orbit-boosting maneuvers, involving five orbit burns around the Earth, spanning a 21-day period. This approach was necessary due to the limitations of the PSLV. Because the PSLV had altitude constraints and couldn't reach the Moon directly, ISRO (Indian Space Research Organisation) had to devise this innovative method. Following the fifth orbit burn, the craft executed its lunar orbit insertion, which was then succeeded by four lunar orbit reduction manoeuvres.

Some of the important objectives of the Chandrayaan-1 were to collect data about the Moon's geology, mineralogy and topography.

India's First Two Missions to the Moon

The mission carried 11 scientific payloads (five Indian instruments and six instruments from other countries).

The Indian payloads were:

- Terrain Mapping Camera (TMC), for high-resolution mapping of the moon.
- Hyper Spectral Imager (HySI), for the mineralogical mapping.
- Lunar Laser Ranging Instrument (LLRI), to give indications about the Moon's topography (height of certain features).
- High Energy X-ray Spectrometer (HEX), for examining radioactive elements on the surface.
- Moon Impact Probe (MIP), which was intentionally crashed into the Moon's south pole. The debris from its impact aided Chandrayaan-1 in its search for lunar water.

The other six payloads included:

- C1XS or X-ray fluorescence spectrometer, from the United Kingdom's Rutherford Appleton laboratory, the European Space Agency (ESA) and ISRO. This was for mineral mapping.
- SARA, the Sub-keV Atom Reflecting Analyser

(ESA instrument) for mapping mineral composition.
- M3, the Moon Mineralogy Mapper (Brown University and NASA) was an imaging spectrometer designed to map the surface mineral composition.
- SIR-2, a near infrared spectrometer (ESA), for mapping the mineral composition using an infrared grating spectrometer.
- Mini-SAR (from NASA) an active Synthetic Aperture Radar system to search for lunar polar ice, water ice. Essentially for the estimation of water content of the Moon's Polar Regions.
- RADOM-7, Radiation Dose Monitor Experiment from the Bulgarian Academy of Sciences for mapping the radiation environment around the Moon[17].

ISRO getting success with such a complicated mission and with minimal budget was appreciated globally. It could be said that this mission has immensely helped to raise the prestige of ISRO as a space agency which is highly professional and can undertake major missions in minimal budget. This mission showed India's inclination for pushing the envelope—especially in space science. Primarily, this was the main reason for

India's First Two Missions to the Moon

India's Moon mission getting conceptualized. India's stated objectives for reaching the Moon were to expand scientific knowledge, upgrade India's technological capabilities and provide challenging opportunities to young scientists working in planetary sciences. Also, India's interest in identifying the presence of Helium-3[18], which is expected to be a better alternative for energy security, was an important objective too.

To support its Moon mission, India established a pair of large antennas for monitoring purposes. This facility, known as the Indian Deep Space Network (IDSN), comprises two powerful dish antennas, one measuring 32 meters and the other 18 meters in diameter. Furthermore, ISRO maintains various ground stations within and outside India under the ISRO Telemetry, Tracking and Command Network (ISTRAC) to provide ground support for Moon and Mars missions.

Chandrayaan-1 mission was responsible for the detection of water (H_2O) and hydroxyl (OH) on the lunar surface. NASA had used the opportunity of India's mission going in the vicinity of the Moon to fly two of its water-hunting instruments onboard. Their Miniature Synthetic Aperture Radar (Mini-SAR) found the patterns of reflected signals from more than 40 polar craters to be consistent with water ice.

NASA had some idea about the possibility of water based on its Clementine Moon-mapping orbiter data. Clementine, the robotic US spacecraft had orbited and observed all regions of the Moon over a two-month period in 1994. It carried out geologic mapping in greater detail than any previous lunar mission; some of its data hinted at the possibility that water exists as ice in craters at the Moon's south pole. However, NASA was keen to have confirmation of this finding. Chandrayaan-1 carried with it an upgraded instrument, NASA's Moon Mineralogical Mapper (M3), that could differentiate between ice, liquid water and water vapor based on how the lunar surface reflected and absorbed infrared light. It was M3 that confirmed our moon hosts water once and for all and found most of it to be concentrated on the poles[19]. Observations from ISRO's Moon Impact Probe (MIP) also gave some idea about the presence of water molecules in the lunar environment.

Salient science results from Chandrayaan-1 (as stated by ISRO in their official website):

1. Discovered the presence of Hydroxyl & water molecules on the lunar surface (M3), lunar exosphere (ChACE on MIP) and sub surface water ice deposits in the base of the craters

of permanent sun shadow region (Mini-SAR).
2. Direct evidence for water (H_2O) in the sunlit ambience from CHACE on MIP of Chandrayaan-I.
3. Reflection of solar wind protons as neutral hydrogen from lunar surface (SARA).
4. Three dimensional conceptualization of many craters of interest and detailed maps of lunar surface features (TMC and LLRI).
5. Detection of potential site (buried lava tube) for future human habitability on the Moon (TMC and HySI). This may provide a safe environment from hazardous radiations, micro-meteoritic impacts, extreme temperatures and dust storms.
6. Evidence of volcanic vent, lava pond and lava channels as recent as 100 million years old inside the Tycho crater central peak (TMC and NASA's LRO data).
7. Enhanced signatures of OH/Water in Compton Belkovich Volcanic Complex, which indicates the endogenic water inherited during the formation of the Moon, now coming out with Volcanism (M3 spectra).[20]
8. The team led by researchers from the University of Hawai'i (UH) published their research in

the journal nature astronomy on 15 September 2023. By analysing the remote sensing data from India's Chandrayaan-1 lunar mission they have found that high energy electrons from Earth may be forming water on the Moon.

Chandrayaan-2

India's second Moon mission was encountering various challenges during the development phase itself. India had signed an agreement with Russia's Federal Space Agency, Roscosmos, for a joint lunar research and exploration mission. This mission was expected to take place during 2014. The mission objectives were as follows:

- To conduct mineralogical and chemical mapping of the lunar surface. This would also help towards unravelling the mystery of evolution of the Moon and Solar System. The Moon surface would be searched for surface or subsurface water ice.
- To study the Moon from the perspective of future landing missions.
- To upgrade the technological base in the country.

Chandrayan-2 mission was to consist of a spacecraft (orbiter) and a landing platform with the Moon rover. The platform was to have a rover-lander system to detach itself after the spacecraft reached its orbit above the Moon and was to land (soft landing) on lunar soil. As per the agreement, India was to be responsible for the orbiter and Russia for the Moon rover and lander system. Unfortunately, Russia was unable to fulfil their commitment owing to problems at their end. When Russia cited its inability to provide the lander even by 2015, India decided to go solo. All this led to a major delay.

Finally, after missing various deadlines, the launch date for Chandrayaan-2 was decided as 15 July 2019. For that day, everything was put in place and finally the countdown had begun. Few minutes were left in the final phase when the launch was called off owing to some glitches. But ISRO was quick in identifying the fault and rectifying it. Within a week everything was rescheduled, and a successful launch occurred on 22 July 2019. Again, like the first Moon mission (Chandrayaan-1), the orbit-raising manoeuvres were carried out and the translunar injection happened on 13 August 2019. The orbiter was placed in the correct orbit and altitude over the Moon's surface by 1 September

2019. On 2 September, the Vikram lander (with the Pragyan rover in its belly) was separated from the spacecraft and it was to perform the soft landing on the moon's surface on 6 September 2019. However, at an altitude of just 2.1 km above the lunar surface there was a trajectory deviation, and unfortunately the lander performed the hard landing thus breaking into pieces. Luckily, the performance of the orbiter was normal. In fact, the projected life of the orbiter was supposed to be one year. But, since much of the fuel was saved, it was estimated that the orbiter would last for seven years from its launch.

ISRO had selected eight scientific instruments for the orbiter. The lander was to have four payloads and two were for the rover. This time, in comparison with the first mission, ISRO was not able to carry any payloads for others (NASA and ESA were interested) owing to weight restrictions. However, a small laser retroreflector from NASA was carried as the lander's payload. It was supposed to be for distance measurement and assisting the rover in navigation on the lunar surface.

Payloads on the orbiter which have been designed and developed by various agencies of ISRO:

- Chandrayaan-2 Large Area Soft X-ray Spectrometer (CLASS) to determine the

elemental composition of the lunar surface.
- Solar X-ray monitor (XSM) for mapping major elements present on the lunar surface.
- Dual Frequency L and S band Synthetic Aperture Radar (DFSAR) for probing the first few meters of the lunar surface for the presence of different constituents, including water ice. The idea was to collect more evidence about the presence of water on the moon's surface and beneath.
- Imaging IR Spectrometer (IIRS) for mapping of lunar surface for the study of minerals, water molecules and hydroxyl present.
- Chandrayaan-2 Atmospheric Compositional Explorer 2 (ChACE-2) to carry out a detailed study of the lunar exosphere.
- Terrain Mapping Camera-2 (TMC-2) for preparing a three-dimensional map essential for studying the lunar mineralogy and geology.
- Radio Anatomy of Moon Bound Hypersensitive Ionosphere and Atmosphere – Dual Frequency Radio Science experiment (RAMBHA-DFRS) for the studying electron density in the Lunar ionosphere.
- Orbiter High Resolution Camera (OHRC) for scouting a hazard-free spot prior to landing.

Basically, for identifying the landing site of the lander.

This mission was planned in a systematic way. The landing site of the lander was at the southern pole of the Moon. This region was specifically chosen due to sunlight never reaching there. Hence, there was a possibility of finding water in its frozen form. India's rover was to operate on the Moon's surface for one lunar day (14 Earth days). All this could have increased the knowledge of the Moon's surface. However, the good part is that the orbiter has already started giving good data inputs and is expected to do so till 2026. Broadly, it could be said that Chandrayaan-2 was a partially successful mission.

Even today, the Chandrayaan-2 orbiter continues to function perfectly. It successfully connected with the Chandrayaan-3 just two days before the Chandrayaan-3 performed soft landing on the Moon's south pole. Post 2019, this orbiter has been continuously providing data related to lunar topography, mineralogy, elemental abundance, lunar exosphere and signatures of water ice. So far, much information had already been gathered and ISRO published a detailed analysis of the scientific findings by each sensor based on the first two years of data[21].

Broadly, the sensors re-established the earlier findings about the presence of water on the lunar surface. Much knowledge about the new lunar craters had been gained. Exploration of the region of south pole provided more detailed information about the craters and boulders underneath the regolith and about the loose deposit comprising the top surface extending up to 3–4 m in depth. This information helped in planning for the Chandrayaan-3 mission. ISRO had used remote sensing techniques for a more detailed understanding of the lunar surface. This data had identified the presence of elements like manganese, sodium and chromium on the Moon.

ISRO is also picking up data on many microflares outside the active region. This is supposed to be new information and would add towards our knowledge regarding solar flares and heating of the solar corona.

Chandrayaan-2 orbiter is currently in a 100 km x 100 km orbit around the Moon. Some other interesting results which are adding to the existing global knowledge include (available at ISRO's official website):

- CHandra's Atmospheric Composition Explorer-2 (CHACE-2) is a quadrupole based neutral mass spectrometer aimed at observing the tenuous lunar exospheric composition from spacecraft altitude. It is known that Argon-40 (Ar-40) is a noble gas in the lunar exosphere. It had been detected by several previous missions, mostly covering the equatorial and low-latitude regions of the Moon. The CHACE-2 not only made observations over the low-latitude regions, but also covered the other latitude regions as well. CHACE-2 observations showed Ar-40 enhancements over certain longitude sectors including KREEP (Potassium, Rare Earth Elements and Phosphorus rich region on the Moon) region and the South Pole Aitken (SPA) terrain. Lunar lobate scarps are relatively small-scale tectonic landforms that are interpreted to be the surface expression of low angle thrust faults. Lobate scarps are believed to be young lunar landforms and are observed both in the mare and highland regions. ISRO's Terrain Mapping Camera (TMC-2) is providing additional information in various regions of the Moon which provides an opportunity to map the features having smaller dimensions such as lobate scarps.

▸ XSM is carrying out broadband spectroscopy

of the Sun from lunar orbit. Currently, XSM is the only X-ray spectrometer in the world which regularly measures the soft X-ray spectrum of the Sun with the highest time cadence. This has yielded very interesting observations of the microflares occurring outside active region as well as elemental abundances in the quiet Sun corona. XSM has also observed number of B-class flares and their analysis has yielded unprecedented observations of variation of the elemental abundances during such flares[22].

4

Chandrayaan-3: ISRO's Moon Supremacy

There is a word called 'selenophile'. Various dictionaries mention the meaning of this word as someone who loves and is obsessed with the moon. The moon has been part of folklore for ages. Broadly, folklore entails traditional beliefs, customs and stories of a community passed through the generations orally. It is said that the moon was one of humankind's first timepieces. Humans found that the moon's face changes nightly and with the regularity of the seasons, making it a reliable marker of time. Science fiction is full of stories on the moon. In 1902, a French science-fiction adventure film was

Chandrayaan-3: ISRO's Moon Supremacy

released called 'A Trip to the Moon' which is based on Jules Verne's 1865 novel *From the Earth to the Moon* and its 1870 sequel *Around the Moon*. In Indian mythology too, one finds numerous references to the moon.

The human fascination towards the moon remains even today. However, as science and technology has progressed, the attraction towards the moon is more scientific rather than a fantasy. As mentioned earlier, ISRO's quest for the Moon witnessed an important high point with ISRO's Chandrayaan-3 mission successfully landing on the lunar surface and commencing the lander (Vikram) and the rover (Pragyan) operations.

It could be said that Chandrayaan-3 was India grandly announcing her goal to ultimately carry out human missions to the Moon in the future. With this successful soft-landing, it could be said that ISRO's love story with the moon has just begun. Particularly, against the backdrop of Chandrayaan-2's failure, this soft-landing of the lander unit on the lunar surface is even more praiseworthy.

Chandrayaan-3 mission was launched on 14 July 2023, and it finally reached the lunar surface on 23 August 2023. Subsequently, after a few hours, the ramp in the lander's belly opened and slowly the rover unit came out. It operated on the Moon's surface for 12 to 14 days and collected a lot of useful information that could come in handy for designing future missions to the Moon. The entire credit goes to the scientific community of ISRO; they did give all in for four years to make this mission a success. It was heartening to witness ISRO beaming with confidence after the success of Chandrayaan-3.

India is the fourth country to successfully undertake a robotic landing on the Moon's surface. Today, major global powers are astonished at the roaring success of India, a State from the Global South, which still gets bracketed as a 'developing State'.

It is important not to look at the success of the Chandrayaan-3 mission in isolation. The success is an outcome of the process which began with Chandrayaan-1 mission. The surprise discovery made by NASA–ISRO about the presence of water on the Moon based on the observations from the Chandrayaan-1 reignited the scientific community all over the world to evolve a major agenda for Moon research. The Chandrayaan-3 mission should be viewed as one such effort in that direction.

Chandrayaan-3: Mission Details

Chandrayaan-3 was a follow-on mission to Chandrayaan-2 to demonstrate end-to-end capability in safe landing and roving on the lunar surface. It was launched by Geosynchronous Satellite Launch Vehicle Mark III (LVM3) from Satish Dhawan Space Centre (SDSC) SHAR, Sriharikota. Chandrayaan-3 consisted of an indigenous lander module (LM), propulsion module (PM) and a rover with an objective of developing and demonstrating new technologies required for interplanetary missions. The lander had proved its capability to soft land at a specified lunar site. For one lunar day (14 Earth days) the rover had carried out in-situ chemical analysis of the lunar surface during its mobility and was known to have travelled a distance of about 100 meters. The lander and the rover had scientific payloads to carry out experiments on the lunar surface, which they did successfully. The main function of the PM was to carry the LM from launch vehicle injection till final lunar 100 km circular polar orbit. It separated from the LM as per the plan. Besides this, the PM had its own scientific payload to undertake various observations.

The mission objectives of Chandrayaan-3 were:

- To demonstrate safe and soft landing on the lunar surface
- To demonstrate rover roving on the moon
- To conduct in-situ scientific experiments

Chandrayaan-3

To achieve the mission objectives, several advanced technologies were present in the lander such as:

- Altimeters: Laser & RF based Altimeters
- Velocimeters: Laser Doppler Velocimeter & Lander Horizontal Velocity Camera
- Inertial Measurement: Laser Gyro based Inertial referencing and Accelerometer package
- Propulsion System: 800N Throttleable Liquid Engines, 58N attitude thrusters & Throttleable Engine Control Electronics
- Navigation, Guidance & Control (NGC): Powered Descent Trajectory design and associate software elements
- Hazard Detection and Avoidance: Lander Hazard Detection & Avoidance Camera and Processing Algorithm
- Landing Leg Mechanism

To demonstrate the aforementioned technologies on Earth conditions, several lander special tests were carried out successfully viz.:

- Integrated Cold Test—for the demonstration of Integrated Sensors & Navigation performance test using helicopter as test platform
- Integrated Hot test—for the demonstration of closed loop performance test with sensors,

actuators and NGC using Tower crane as test platform
- Lander Leg mechanism performance test on a lunar simulant test bed simulating different touch down conditions[23]

For Reaching the Moon

The launch vehicle used for this mission was LVM-3. When ISRO is launching satellites into low earth orbit, they use this version of GSLV-Mk-3. A few months back, LVM-3 was used for undertaking two commercial launches for the UK-based company OneWeb, which provides satellite-based internet facilities. During each mission, 36 satellites were launched into low Earth orbit. Hence, the reliability of the vehicle was already established. Still, the Moon mission was a different game altogether. LVM3 is configured as a three stage vehicle with two solid strap-on motors (S200), one liquid core stage (L110) and a high thrust cryogenic upper stage (C25). Interestingly, while using LVM-3 for Chandrayaan-3, ISRO smartly took the opportunity to use the capacity of this vehicle to undertake future human missions (Gaganyaan). During this mission, ISRO used human-rated solid strap-on motors. Possibly, the success of this launch gave confidence for undertaking future unmanned and manned missions for the Gaganyaan project using the same rocket.

On 14 July 2023, ISRO's LVM-3 had put the Chandrayaan-3 module at an altitude of 180 km, and for reaching the Moon it had to travel about 3,84,400 km up in the sky. To do this, ISRO undertook around five Earth orbit-raising manoeuvres. This allowed the craft to reach about 1,00,000 km above the Earth's surface. Subsequently, ISRO took a slingshot to make it reach a trajectory to the Moon. After it established itself in a lunar orbit, further activity of reducing the altitude started and the craft successfully performed the soft-landing on the Moon's surface.

Major space powers like the US and Russia can reach the Moon within four to five days after the launch. However ISRO, not having a strong rocket launcher system to directly enter the Moon's orbit, needed 40 days to reach there. ISRO understood how to get over this limitation and put in place an innovative process to travel this distance of almost 4,00,000 km. During the Chandrayaan-2 mission too, they had used the same process. Still, the entire process of reaching the Moon is complicated and anything could have gone wrong at any point in time. The process involved three main stages—Earth

Chandrayaan-3: ISRO's Moon Supremacy

orbit manoeuvres, translunar injection and lunar orbit manoeuvres. These stages were completed successfully and the lander separation from the PM was carried out as planned. Subsequently, the lander system was made to enter an orbit closer to the Moon and finally soft landing took place. The following diagram indicates details of this process:

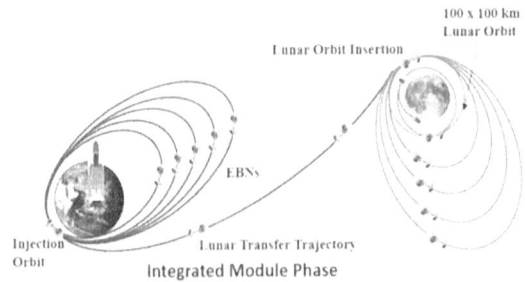

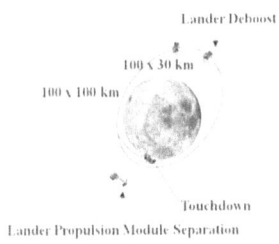

Integrated Module Phase

Source: ISRO

Chandrayaan-3

Chandrayaan-3—Mission Profile

The following description explains how the orbits were raised to approach the Moon and, subsequently, how the spacecraft's height was lowered to reach the lunar surface. ISTRAC in Bengaluru was responsible for executing all these manoeuvres and injection with precision[24].

On 14 July 2023, LVM-3 had put the Chandrayaan-3 module at an altitude of 180 km, in an Elliptic Parking Orbit (EPO) of size 36,500 km × 180 km. The first orbit raising manoeuvre was performed on 15 July and craft was placed in 41,762 km × 173 km orbit. The second orbit raising manoeuvre was performed on 16 July and craft was placed in 41,603 km × 226 km orbit. This was followed by third manoeuvre on 18 July which put the craft in 51,400 km × 228 km orbit. The 20 July orbit raising manoeuvre had placed the craft in 71,351 km × 233 km orbit. The fifth and final orbit raising manoeuvre was performed on 25 July putting the spacecraft in an orbit of 127,609 km × 236 km.

On 1 August 2023, the translunar injection was performed to propel Chandrayaan-3 from Earth's orbit and to begin its journey towards the Moon. Much precision is required to undertake this operation

Chandrayaan-3: ISRO's Moon Supremacy

to allow the spacecraft to break free from Earth's gravitational force. On 5 August, the Chandrayaan-3 mission achieved a crucial milestone with the successful completion of the Lunar Orbit Insertion (LOI). The manoeuvre resulted in an orbit of 164 km × 18,074 km. After the LOI, the mission was to undergo four more planned orbital manoeuvres to position the spacecraft at an approximate distance of 100 km from the Moon's surface. On 6 August, the spacecraft was in a 170 km × 4,313 km orbit around the moon. On 9 August, the orbit was reduced to 174 km × 1,437 km. The spacecraft was in 151 km × 179 km orbit on 14 August. Finally, on 16 August, the spacecraft was in an orbit of 153 km × 163 km. On 17 August, the LM successfully separated from the PM and de-boosting was carried out on 18 August. The LM then, was in 113 km × 157 km orbit around the Moon. The second de-boosting on 20 August, took the spacecraft in 25 km × 134 km orbit. This was followed by the powered descent on 23 August 2023, around 5.45 p.m., and the spacecraft was safely landed on the lunar surface.

The most critical part of the entire mission was the phase of final descent on the lunar surface. This phase was to last for around 15 to 18 minutes. It was called as 'period of terror' because the entire process of descent and landing was to happen in an autonomous mode. Ground control had no say in these activities. ISRO scientists had fed the system's software with multiple options and the spacecraft was trained to even handle last minute glitches (if any). This time, ISRO was wiser as they had already learnt 'what not to do' from Chandrayaan-2 mission's failure. Chandrayaan-3's powered descent had four broad phases: Rough Braking Phase; Attitude Hold Phase; Fine Braking Phase and Terminal Descent Phase.

One critical part of the landing involved reducing the lander's horizontal velocity from a range of 1.68 km/sec (more than 6,000 km/h) at a height of 30 km from the lunar surface to almost zero for a soft landing at the designated site about 70 degrees south latitude. Chandrayaan-3 was tilted almost 90 degrees when the last phase operations were to begin. The criticality involved was to make the system vertical for a landing during the process of descent. ISRO was able to achieve a perfect landing. The entire efforts were to ensure that during the last 15 min or so, everything should function as planned. Required speed reduction as envisaged did happen. The lander system had eight camaras which were continuously taking pictures of the pre-decided landing site. The orbiter launched during the Chandrayaan-2 mission

Chandrayaan-3: ISRO's Moon Supremacy

had already established contact with the lander unit and was also providing pictures of the landing site taken from 100 km above the Moon surface. The system was comparing the pictures in real-time and was deciding on the feasibility of the landing site to undertake landing. The plan was: if the landing site was not found suitable, then choose an alternative site 150 meters away for landing. ISRO had ensured that even if the lander unit had not landed perfectly, but with 10 to 12 degrees of tilt, still the system would have performed normally.

The landing site was chosen near the south pole of the Moon at 70 degrees latitude (4 km x 2.4 km, 69.367621°S, 32.348126°E). This craftily handled landing is an exceptional achievement since the south pole area has large craters and uneven surface. Various earlier Moon missions had landed close to the equator, the region being safe for landing. Also, terrain and temperatures were helpful for landing. More importantly, assured sunlight presence was helpful for operation of solar-powered instruments. In comparison, the south pole was a challenge. ISRO chose to land not exactly on the south pole but close to 70 degrees south latitude. The purpose of ISRO's mission was to know more about the water ice deposits on the Moon and hence the south pole was the best option. But temperatures in this region can go as low as -230°C which made it imperative to ensure that their craft and sensors would withstand such low temperatures. For ISRO, choosing a place and managing a perfect landing over there was a real test. They were not keen to go to a completely dark region because they wanted the presence of sunlight for operating their instruments on the lunar surface and at the same time wanted to be close to the area (without sunlight) where possibility of water is there.

There were two modes of communication with the stations on the ground. The lander could communicate with the orbiter (Chandrayaan-2) and then the orbiter would pass on the information to the ground station. Other approach was that the lander could directly connect with the ground station. In case of the rover, there was only one communication link: it could only get in touch with the lander and vice versa.

Communication is an indispensable part of every deep space mission. For any space agency, when the spacecraft is travelling towards its target, it is important to keep the spacecraft visible and connected with the ground stations. Also, continuous connectivity is required even after the spacecraft has been declared operational. The geographical location of ground stations is vital in this regard. The ground stations located on Indian soil were not able to keep a continuous track of the spacecraft. Over the years, ISRO has established their own ground stations in different parts of the world to keep a track of their various satellite missions. However, for deep space missions, the requirements are different. The size of the antenna was becoming an issue.

Traditional antennas are built to receive and transmit electromagnetic waves, which travel close to the speed of light. But electromagnetic waves have a relatively long wavelength. Hence antennas were required to maintain a certain size to work efficiently with electromagnetic radiation. NASA has the biggest DSN with a 70 m (230 ft) diameter antenna. Such antennas can track spacecraft travelling tens of billions of kilometres from Earth. For the Chandrayaan-3 mission, ISRO had taken help from NASA and ESA for coordination in deep space. Three ground stations of ESA located in Kourou (French Guiana, 15 m antenna), Goonhilly (United Kingdom, 32 m antenna) and New Norcia (Western Australia, 35 m antenna) were of great help. NASA's DSN did provide the telemetry and tracking coverage during the powered descent phase from their stations DSS-36 and DSS-34 located at Canberra Deep Space Communications Complex followed by DSS-65 at Madrid Deep Space

Communications Complex[25]. ISRO has its own IDSN with 18 m and 32 m antennas.

Mission Payloads[26]

Lander Payloads: Chandra's Surface Thermophysical Experiment (ChaSTE) to measure the thermal conductivity and temperature, Instrument for Lunar Seismic Activity (ILSA) for measuring the seismicity around the landing site, Langmuir Probe to estimate the plasma density and its variations. A passive Laser Retroreflector Array from NASA was accommodated for lunar laser ranging studies.

Rover Payloads: Alpha Particle X-ray Spectrometer (APXS) and Laser Induced Breakdown Spectroscope (LIBS) for deriving the elemental composition in the vicinity of landing site.

The objectives of scientific payloads planned on Chandrayaan-3 lander module and rover are provided below:

Table 1
Objectives of Lander Payloads

Sl. No	Lander Payloads	Objectives	
1.	Radio Anatomy of Moon Bound Hypersensitive ionosphere and Atmosphere (RAMBHA)	Langmuir probe (LP)	To measure the near surface plasma (ions and electrons) density and its changes with time
2.	Chandra's Surface Thermo physical Experiment (ChaSTE)	To carry out the measurements of thermal properties of lunar surface near polar region.	
3.	Instrument for Lunar Seismic Activity (ILSA)	To measure seismicity around the landing site and delineating the structure of the lunar crust and mantle.	
4.	LASER Retroreflector Array (LRA)	It is a passive experiment to understand the dynamics of Moon system.	

Table 2
Objectives of Rover Payloads

Sl. No	Rover Payloads	Objectives
1.	LASER Induced Breakdown Spectroscope (LIBS)	Qualitative and quantitative elemental analysis and to derive the chemical Composition and infer mineralogical composition to further our understanding of Lunar-surface.
2.	Alpha Particle X-ray Spectrometer (APXS)	To determine the elemental composition (Mg, Al, Si, K, Ca, Ti, Fe) of Lunar soil and rocks around the lunar landing site.

Table 3
Objectives of Propulsion Module Payload

Sl. No	Propulsion Module Payload	Objectives
1.	Spectro-polarimetry of Habitable Planet Earth (SHAPE)	Future discoveries of smaller planets in reflected light would allow us to probe into variety of Exo-planets which would qualify for habitability (or for presence of life).

Chandrayaan-3 before launch at Sriharikota

Top: Preparing for a long journey
Bottom: Moving towards the destination

Kangaroo on the Moon: Vikram lander and Pragyan rover

Top: No substitute to hard work
Bottom: A majestic launch

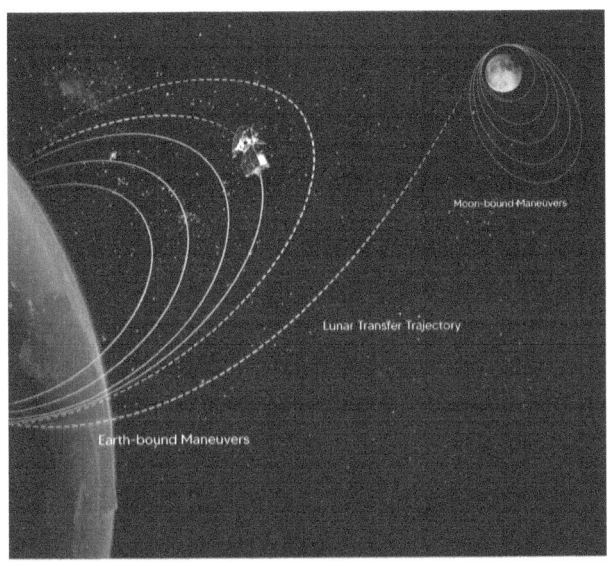

Top: A step-by-step journey. Chandrayaan-3 route in sketches
Bottom: Chandrayaan-3 preparing for a perfect soft-landing

ChaSTE　　　　ILSA

Top: Vikram as seen by Pragyan
Bottom: In search of both far and near! Lunar far side area as imaged from a camera onboard Chandrayaan-3.

Top: Chandrayaan-3: 3,84,400 km away from Earth
Bottom: Pragyan is ready for a Moon Walk!

A long way to go. The path retraced by the Chandrayaan-3 rover

Chandrayaan-3: ISRO's Moon Supremacy

By 2 September 2023, Chandrayaan-3's rover Pragyan completed its assignments, and it was safely parked and set into sleep mode. APXS and LIBS payloads were turned off. At the time of going into hibernation, the Rover's battery was fully charged, and the solar panel was oriented to receive the light at the next sunrise on 22 September (the receiver was kept on). ISRO had planned this mission to last for one lunar day and the mission had already accomplished this task. However, there was some possibility that, if the lander and rover unit could withstand the low lunar temperatures during lunar night (could reach -230°C), then the mission life could get extended for another lunar day.

During this one lunar day stay on the Moon's surface, the Rover did encounter a challenging situation. On 27 August, the Rover came across a 4 m diameter crater positioned 3 m ahead of its location. The Rover was commanded to retrace its path and it headed in a new direction. ISRO declared some interesting observations taken by the lander and rover system. The ChaSTE instrument on the Vikram lander provided the first profile of lunar surface temperature variation from the vicinity of the Moon's south pole. To understand the thermal behaviour of the Moon's surface, ISRO's probe had taken temperatures at the surface and 10 cm beneath the surface. This was specifically designed for this purpose and was fitted with 10 individual temperature sensors. Chandrayaan-3 probe showed a 50°C difference between the Moon's surface and around 10 cm below. On 27 August, ISRO released a graph of the temperature variation between the lunar surface and a point around 8 cm below based on ChaSTE observations. However, these were some initial observations and based on more data, ISRO would conduct a detailed study on this aspect.

Pragyan rover confirmed the presence of sulphur (S) on the surface of the Moon. ISRO also announced that the Rover detected elements like aluminium (Al), calcium (Ca), iron (Fe), chromium (Cr), titanium (Ti), manganese (Mn), silicon (Si) and oxygen (O).

ISRO conducted an interesting experiment just before the system went to sleep. The lander performed a short 'hop'. For this purpose, briefly, the thrusters were fired. The lander moved by about 16 inches (40 cm) and came much closer to the already sleeping Pragyan rover. This hop may be seen as a test for a future sample return mission. For any sample return mission, the craft must first land on the lunar surface, collect the samples and then launch itself towards the Earth from the surface of the Moon. This one small experiment (which was successful) could go a long way in teaching ISRO how to 'take-off' from the lunar surface.

5

Learning from Chandrayaan-2's Failure

Chandrayaan-2 mission had to witness failure when its lander and rover units were descending towards the surface of the Moon. Every activity was going as per plan, but finally when the lander unit was really close to the lunar surface, it crashed (hard landed). It was a rude shock for ISRO's prestige. However, the scientists at ISRO were fully aware that space is an unforgiving medium and failures are part of the game. There is a general notion that science progresses more because of failures.

Initially, despite realizing that the lander unit had performed a hard landing, ISRO kept on trying

Learning from Chandrayaan-2's Failure

to establish contact with the system thinking that possibly everything was not lost, and that the system may respond despite it hitting the lunar surface.

Technically, there was a faint possibility for a response. However, there was no harm in trying. ISRO is an organization known for transparency but somehow it was not forthcoming with the information at that moment. The Vikram lander's powered manoeuvres to land on the Moon worked just fine until it was 2.1 km above the surface. Naturally, ISRO almost had data till final stages of the mission and was aware of the landing zone.

Hence, by extrapolating, they could have got some idea about the possible location of the fall of the lander on the lunar surface. They could have tasked Chandrayaan-2 orbiter for probing an approximate search area. Another option was NASA's Lunar Reconnaissance Orbiter (LRO), which could have helped identify the status of Vikram provided ISRO shared some data with them. Interestingly, Shanmuga Subramanian, a Chennai-based engineer was continuously monitoring the LRO data out of curiosity. He identified some traces and thought that it could be a piece of Vikram. He contacted the LRO team for confirmation. The team expanded on the clue given by him and confirmed that it was the wreckage of Vikram. However, ISRO remained silent on this finding and, subsequently, declared that their own orbiter had identified the location of Vikram on the lunar surface[27]. Normally, ISRO makes their failure report public, however in the case of Chandrayaan-2, they remained tight-lipped until recently. A few months before the launching of Chandrayaan-3 mission, present ISRO chairman Mr S. Somanath spoke at length about the reasons for the failure of Chandrayaan-2. His revelations to various media platforms have shed light on what possibly went wrong and how ISRO has worked to overcome those challenges for Chandrayaan3 mission.

It would be of interest to know the entire mission profile[28] of Chandrayaan-2 just to appreciate the mission progress and some problems it faced from the beginning itself. It is difficult to say whether the failure was the result of any specific reason or happened owing to the combination of all these problems.

This mission had two major parts: the orbiter and a lander-rover system. The orbiter was a 2,379 kg satellite while the lander's weight was 1,471 kg and the rover's 27 kg.

Chandrayaan-2, since the launch at 6 August, underwent five orbit-raising manoeuvres which increased its apogee to around 1,50,000 km.

Learning from Chandrayaan-2's Failure

Subsequently, on 14 August, translunar insertion took place, followed by the Lunar Orbit Insertion (LOI) manoeuvre on 20 August. Once in orbit around the Moon, five manoeuvres lowered its orbit. After the separation of the lander took place on 2 September, it successfully performed deorbiting manoeuvres on third and fourth September, and the final landing was attempted on 7 September. Some available details about this travel are presented in form of a table below:[29]

Table 4
Locations During Travel to Moon

Date	Orbit-raising manoeuvre	Proposed	Actual	Deviation
24 July	First	230 × 45,162 km	230 × 45,163 km	1 km
26 July	Second	250 × 54,689 km	251 × 54,829 km	1 × 140 km
29 July	Third	268 × 71,558 km	276 × 71,792 km	8 × 234 km
2 Aug	Fourth	248 × 90,229 km	277 × 89,472 km	29 × 757 km

Chandrayaan-3

6 Aug	Fifth (final)	221 × 143,585 km	276 × 142,975 km	55 × 610 km
14 Aug	Trans	Lunar	Insertion	(TLI)
20 Aug	Moon Entry	118 × 18,078 km	114 × 18,072 km	4 × 6 km
21 Aug	Second Lunar orbit manoeuvre	121 × 4303 km	118 × 4412 km	3 × 109 km
28 Aug	Third	178 × 1,411 km	179 × 1412 km	1 × 1 km
30 Aug	Fourth	126 × 164 km	124 × 164 km	2 km
1 Sep	Fifth	114 × 128 km	119 × 127 km	-5 × 1 km
2 Sep	Lander	separated	from	Orbiter
2 Sep			119 × 127 km	
	De-orbiting	Manoeuvre		
3 Sep	First	109 × 120 km	104 × 128 km	5 × -8 km
4 Sep	Second	36 × 110 km	35 × 101 km	-1 × 9 km
7 Sep	Hard	Landing	Occurred	

Source: https://www.thespacereview.com/article/3793/1

Learning from Chandrayaan-2's Failure

This table indicates that on a few occasions there were significant deviations between proposed and actual orbits. But they could be considered less significant from the point of view of the mission's health, if you consider the amount of the distance the orbiter was travelling. ISRO did not, at the time, make any official comments about these deviations.

India's Moon Missions

For any planetary exploration, landing on the surface of a planet is the most challenging part. Former ISRO chief K. Sivan had pronounced on various occasions that the last part of the mission which involves the controlled (powered) landing is terrifying! This involved the travel of the lander for the last 35 km. This trip was to take 15 minutes and became known as the '15 minutes of terror.'[30]

Everything was supposed to happen in autonomous mode during this 15 minute period of descent. ISRO had no previous experience of designing and handling missions which were fully in autonomous mode. A deviation in the trajectory, 12 minutes into the descent, was visible. Obviously, the system had malfunctioned during the autonomous mode. Possibly, the velocity was higher than required at that height.

There was also a view that the problem could have involved some kind of glitch in the functioning of engines used in the descent. Possibly, the system was not trained to go in safe mode anticipating the problems. Also, there was a possibility that the lander had gone too close to the landing site for the system to take any corrective measures.

It is not the purpose of this book to undertake any detailed failure assessment. Based on available information within a few days after the mission, it became clear that the system (lander unit) was not trained adequately to identify the fault and take corrective measures. The strength of autonomy of any system depends on the basic design factors. The design is an outcome of what the scientists and technologists anticipate as possible anomalies and feed them in the system. Normally, autonomous systems operate in complex and open-ended environments with high levels of independence. They are expected to learn and reason themselves. In addition, they are designed in such a way that they would identify the unforeseen changes in time and respond accordingly. There was no idea regarding the nature and quality of artificial intelligence (AI) involved in the entire process of the landing of Vikram. The Rover was supposed to be a totally AI-based product, but it never got an

Learning from Chandrayaan-2's Failure

opportunity to function. Was there any problem with AI design? It could be difficult to quantify the role, if any, played by AI during the last few minutes of the mission. AI works on data inputs and for such unique missions, it is unlikely that required data would have been available in abundance. Normally, deep space is a data sparse region and the AI based models might not function accurately if they depend only on real-time data. Such models need to be trained based on various possibilities.

After the failure of Chandrayaan-2 to successfully perform soft-landing on the lunar surface, some talk about the possibility of lunar gravity impacting the success of the mission started happening. The Moon is 'lumpy' due to uneven mass distribution under its surface. Compared to Earth, the Moon has less gravity, very little atmosphere and lots of dust; all this could have posed some challenges for landing. Since the Moon's location continuously changes due to its orbital motion, the intersection of its path with Chandrayaan-2 should have been predicted sufficiently in advance with a higher level of accuracy[31]. There is no idea if this calculation was done during the Chandrayaan-2 mission or not. The acceleration due to gravity on the surface of the Moon is approximately 1.625 m/s², about 16.6 per cent of that on Earth's surface. Earth's gravity is 9.807 m/s² and the Moon's gravity is one–sixth of this. It is known that for any planet, gravity is not constant everywhere and reduces towards the pole. It has been estimated that on the lunar surface, the variation in gravitational acceleration is about 0.0253 m/s². However, some opine that the Moon's gravity may not have been a problem for landing in case of Chandrayaan-2. In addition, deep space communication could also have been a challenge since ISRO had limited experience in this field.

Chandrayaan-2 mission had five engines which were used to reduce the velocity (called retardation). These engines developed higher thrust. When such higher thrust was happening, the errors on account of this differential were accumulated over some period. All the accumulated errors were higher than the expectations of the system designers. The software was not tuned to handle the lander moving at higher speed and fast turning. Also, for landing, a very small sized zone (500 m × 500 m) was selected thus giving the lander less flexibility for last minute required deviation. As per Mr Somanath, while designing for Chandrayaan-3, all these issues were kept in mind. On the lander unit, the number of engines were reduced from five to four. For a landing site, an area of 4.2

km (along the track) × 2.5 km (width) was identified. More importantly, ISRO changed the context of designing structures. Instead of success-based design, ISRO opted for failure-based design. The idea was to think rationally about what can fail and how to protect it. Various possible failure scenarios were calculated and programmed. Lots of simulations were carried out. Various possible sensor failures, engine failures, algorithm failures and calculation failures were thought of and corrective measures were fed into the software. Additional solar panels on other surfaces of the lander were mounted to ensure that it generates power no matter how it lands[32].

Lessons Learnt

In general, it could be said that ISRO undertook a detailed assessment for the reasons of failure during the Chandrayaan-2 mission. While planning for Chandrayaan-3 mission, they had taken numerous precautions to ensure that no glitches took place. Very thoughtful software and hardware changes were made based on results of various equipment testings and simulations. Various scenarios were generated to understand wider dispersions which may happen. Algorithms were built based on diverse possibilities leading to problems in the mission, and various alternative solutions were worked out and incorporated into the system. The spacecraft had more fuel for any last-minute challenges. Also, there was no orbiter requirement for the Chandrayaan-3 mission. This allowed ISRO to make the lander more robust. Based on the last mission's experience, ISRO decided to have the lander with strong legs. The Chandrayaan-3 lander weighed 252 kg more than the earlier mission. ISRO had identified all 'known unknowns' and designed the new system accordingly.

6

Industry and People

Public and Private Industry Participation

During March 2019, the Government of India had announced the formation of NewSpace India Limited, a public sector undertaking which is responsible for producing, assembling and integrating the launch vehicle with the help of industry consortium. Subsequently, Indian National Space Promotion and Authorisation Centre (IN-SPACe) has been established by the government to perform the role of a regulator for the space exploration industry. In addition, there are two private associations which help connect private space industry with the government agencies:

- Indian Space Association (ISpA), an apex non-profit industry body, set up exclusively for the successful collaborative development of the private space industry in India.
- SatCom Industry Association (SIA-India) is a non-profit organization representing the interests of the space industry in India and has members which include satellite operators, manufacturers, suppliers, startups, academic institutions and law firms.

In recent times, these organizations are found playing an important role towards connecting industries with ISRO. They also conduct conferences, publish specific reports and come out with routine publications like monthly reports. Some of their recent publications and announcements provide some details about the pivotal role of Indian private industry in the Chandrayaan-3 project.

It is also important to mention that the Indian public sector agency Hindustan Aeronautics Limited (HAL) supplied critical components to National Aerospace Laboratories for launch vehicle testing. Way back during 1983[33], ISRO entered into a long-term cooperation agreement with HAL through a MoU. Aerospace division of HAL was established to fabricate

Industry and People

light alloy structures and tankages for satellite launch vehicles & spacecraft related programmes of ISRO. Today, HAL supports all the major and challenging space programmes of India.

Another Government of India enterprise which is under Ministry of Defence called Mishra Dhatu Nigam Limited (MIDHANI) Hyderabad[34] has a partnership with ISRO which spans over four decades. MIDHANI products of cobalt base alloys, nickel base alloys, titanium alloys, and special steels and investment castings are used in LVM3-M4 liquid engines, nozzles for liquid stages, gas bottles, thrusters, cryogenic upper stage components, rocket motor casing and propellant tanks which carry Chandrayaan-3 payloads. The ultra high strength steel strips made by MIDHANI are used in the propulsion module of lander separator band. They have also provided titanium rings, bars and blocks for Radio Anatomy of Moon Bound Hypersensitive ionosphere and Atmosphere (RAMBHA) and ChaSTE payloads of the lander.

The Ministry for Micro, Small and Medium Enterprises (MSME)[35] of the Government of India's Bhubaneswar Tool Room had manufactured about 54,000 aerospace components of 437 varieties for the mission. Also, Institute for Design of Electrical Measuring Instruments (IDEMI) Mumbai, is known to have played an important role in manufacturing parts for Chandrayaan-3. Semiconductor Laboratory (SCL) Mohali, a research institute under the Ministry of Electronics and Information Technology, Government of India is known to have played a role towards fabrication of the Vikram Processor and Complementary Metal Oxide Semiconductor (CMOS) camera configurator for navigation and imaging. Kerala Minerals & Metals Ltd. (KMML), a company fully owned and managed by the Government of Kerala, is known to have supplied titanium sponge alloys for critical components. Another public enterprise from Kerala called Kerala State Electronics Development Corporation Limited (KELTRON) helped in the production of electronic power modules and the test and evaluation system.

Following are details of the role played by some Indian private agencies in making Chandrayaan-3 a success:

1. Ananth Technologies made significant contributions to India's launch vehicle technology, particularly in the development of avionics packages. Their involvement includes on-board computers, navigation systems, control electronics, telemetry and power systems. Additionally, they have played a key role in creating interface packages, power

Industry and People

switching modules, relay and balancing units and other critical components for recent launches. Ananth Technologies has also been integral in the development of major satellite systems for the Chandrayaan-3 programme encompassing telemetry, telecommand, power management systems and DC-DC converters for the mission.

2. Walchandnagar Industries manufactured components of the lunar mission vehicle, the first-stage booster and flex nozzle control tanks with a height of 80 ft and diameter of more than 12 ft.
3. Godrej Aerospace has played a significant role in India's space missions, including the LVM3 launch vehicle and Chandrayaan missions. They have been instrumental in the production of key components such as engines, thrust chambers, and critical parts. These contributions have greatly contributed to the advancement of India's space exploration capabilities showcasing the growth and expertise of the country's space industry.
4. Centum Electronics played a crucial role in the LVM-3 M4 mission by supplying over 200 mission-critical modules and subsystems. Notably, their Ceramic Servo Accelerometers, equipped with advanced electronics, were a key component of the Laser Inertial Referencing & Accelerometer

Package (LIRAP). This technology significantly enhanced accuracy during both launcher and lander operations, achieving unparalleled precision in micro-'g' sensing.

5. Larsen & Toubro (L&T) provided essential components for India's lunar mission, Chandrayaan-3. These components, including the 'middle segment and nozzle bucket flange,' were produced at their Powai facility, while the ground and flight umbilical plates were manufactured at their aerospace facility in Coimbatore.
6. Bharat Heavy Electricals Ltd (BHEL) supplied batteries for the mission.
7. MTAR Technologies contributed parts for the lander's propulsion system and the rover's navigation system.
8. Himson Industrial Ceramic supplied components capable of withstanding extreme temperatures.
9. Sri Venkateswara Aerospace constructed Vikram's legs and the Rover's chassis.
10. Paras Defence and Space Technologies contributed to the navigation system.
11. Omnipresent Robot Technologies developed the perception navigation software used by the Pragyaan rover.
12. Tata Consulting Engineers Limited (TCE)

engineered critical systems and sub-systems, including the solid propellant plant, the vehicle assembly building, and the mobile launch pedestal.
13. Kortas Industries Pvt Ltd (ADTL Subsidiary) contributed several subassemblies for the S200 booster stage, the L110 core stage, and the C25 cryogenic stage, including components for the CE20 cryoengine of the LVM3 launch vehicle.
14. Vajra Rubber Products supplied the S200 thrust vector control flex seal for the LVM3 rocket.

People

A career in science is about exploiting several processes and methods to reach conclusions. It is said that the scientific community needs to observe, think and hypothesize problems. They need to experiment with it, provided it is not a theoretical problem. Mostly to confirm their findings, they are required to repeat tests a few times more till a definitive conclusion is reached. Scientists work tirelessly on the problem in front of them. However, some problems require a group of people working together. Modern-day science is also about agencies undertaking mega projects. Such projects demand expertise from different branches of science. Projects like Chandrayaan-3 could be viewed

Chandrayaan-3

as multidisciplinary projects. ISRO has various subject specific laboratories in the country. Reaching the Moon is all about science and technology. Obviously, various units of ISRO were working on this project simultaneously. Some names are mentioned below who are known to have played a significant role towards making Chandrayaan-3 a success. It is important to mention here that in any scientific laboratory countless individuals who are responsible for their part of the job, usually do not get the limelight although they play an important role towards the advancements in science and technology. They are the driving force behind the success of any major project. There could be some individuals like this in ISRO too, whose names may not have reached the media. Normally, such things do happen in big projects, not necessarily by design, but mostly owing to oversight. It is important to mention here that everyone involved (directly or indirectly) in this project is worthy of acknowledgement.

Scientists behind the success of Chandrayaan-3[36]:

1. S. Somanath, chairman of ISRO since 2022. A mechanical engineer by training, he has played a major role in the past towards designing the rockets. He was the most visible face of ISRO during the

Industry and People

mission. An articulate individual, he deserves all the credit for explaining the past, present and future of the mission in a lucid language.

2. P Veeramuthuvel, Project Director, Chandraayan-3. An IITian known for his technical skills, he was working on this mission since 2019.
3. S. Unnikrishnan Nair, Director, Vikram Sarabhai Space Centre. He was in charge of building the launch vehicle. He has a PhD in Mechanical Engineering from IIT Madras.
4. B. N. Ramakrishna, Director, ISTRAC, head of deep space communications. He is an expert in the areas of navigation and orbit determination of spacecrafts.
5. S. Mohana Kumar, Mission Director, Chandrayaan-3. He was closely associated with various LVM 3 missions.
6. M Sankaran, Director, U R Rao Satellite Centre. He was involved in design, development and realization of all ISRO satellites including Chandrayaan-3. He was also responsible for testing the strength of the Vikram Lander.
7. Kalpana K, Deputy Project Director, Chandrayaan-3. An engineer who has also worked in Chandrayaan-2 and Mars mission projects.
8. Muthayya Vanitha, the first female project

director at ISRO and the first woman to lead an interplanetary mission (Chandrayaan-3). She is an electronics system engineer and was the Deputy Director for the Chandrayaan-3 mission.

9. Ritu Karidhal Srivastava, a senior scientist from ISRO and also known as the 'Indian Rocket Woman', came to fame with her involvement in the 2013–14 Mars mission and was also associated with Chandrayaan-3 mission.

7
Global Moon Agenda[37]

Till date, during the period of 1969–72, six manned missions (12 astronauts, Apollo programme of the US) have landed on the Moon. Subsequently, nothing much happened in the domain of the Moon barring a few robotic missions in the twentieth century. However, twenty-first century is witnessing a rising global interest towards lunar exploration. This chapter takes a broad overview of the missions to the Moon that have been undertaken during the last two decades or so and some that are in the pipeline. In some cases, limited historical backdrop has been provided just to highlight the progressions (if any) in the programmes.

In the twenty-first century, the biggest moon

agendas are that of China and the US. Both these States appear to be in competition to become the first twenty-first century State to put humans on the Moon. Beside these two States, there are other States who are also trying to reach the moon including Russia, Japan, South Korea, North Korea and Israel.

The US

Post 1972, the US did undertake a few Moon missions mostly as flyby and orbiter missions. Presently, the US has undertaken one of the most ambitious Moon programmes of this century called Artemis Program, for which NASA is collaborating with ESA and the space agencies of Japan and Canada. Around 2017, the US administration called on NASA to lead an innovative and sustainable programme of exploration of our Solar System. The idea is to have commercial and international partners as a part of this programme and initially focus on return of humans to the Moon. Till date, around 28 States including India have joined this idea by signing Artemis Accords, which is NASA's articulation about how the world should handle technological and policy matters for planning and executing missions to other planets.

At present, NASA is working on a programme for

the return of humans to the Moon. NASA has devised Artemis's three-part plan. They are undertaking various tests to validate the human rating of their Orion spacecraft, testing the heat shield and other technologies. In this connection, NASA launched Artemis 1 (the Orion spacecraft), an uncrewed Moon orbiting mission on 16 November 2022 and the programme got completed on 11 December 2022. This mission was uncrewed to test the safety of the Space Launch System (SLS rocket), and the Orion capsule's ability to reach the Moon, perform in lunar orbit and return to Earth for an ocean splashdown. The SLS rocket had carried 10 CubeSats into space to perform experiments and technology demonstrations. This mission came within roughly 128 km of the lunar surface (closest approach on 5 December 2022) and realized a maximum distance from Earth of 432,210 km. With this, the Orion broke the record for the farthest distance from Earth travelled by an Earth-returning human-rated spacecraft. The earlier record was held by the Apollo 13 mission, which is known to have reached 4,00,171 km.

Artemis 1 mission was unlike the robotic landing mission by India. Within 89 minutes after lift-off, the craft had carried out the translunar injection manoeuvre. In the case of Chandrayaan-3, on the nineteenth day after the launch, translunar injection was carried out, while the Lunar Orbit Insertion (LOI) was carried out on the twenty-third day after the launch. In the case of the US mission, on the sixth day after the launch, Orion had completed one flyby of the Moon and subsequently entered a distant retrograde orbit for six days thus completing a second flyby of the Moon on the twenty-first day. The Orion spacecraft returned and re-entered Earth's atmosphere and splashed down into the Pacific Ocean on 11 December. The comparison highlights the limitations of India's Moon programme. It needs to be noted that due to absence of stronger rockets to undertake such major missions, India is depending on taking a longer route to reach the Moon. The main problem over here is that India cannot carry more weight to such a long distance. This limits India's capacity to carry heavier and a greater number of payloads. Hence, India's missions generally remain as technology demonstration missions. Still, credit to ISRO for not letting these limitations get in the way of scientific advancement. India needs to figure out way to ensure that they can undertake more scientific experimentation in future missions.

Artemis 2 is slated to carry four astronauts; the Orion capsule will take the crew farther from Earth

Global Moon Agenda

than humans have ever travelled before. This will be a lunar flyby mission with the crew scheduled to complete the entire mission in around 10 days' time and return to Earth. The focus of this mission will be to evaluate the spacecraft's systems while carrying humans. Finally, Artemis 3 will realize the dream of the next man and first woman to step onto the lunar surface. This mission will take place depending on the success of Artemis 2. During this mission, the astronauts will shoot towards the moon using the lunar lander to lower two people to the moon's south polar region. They are expected to be on the moon for around a week. NASA has awarded $45.5 million to 11 US companies, including Elon Musk's SpaceX and Jeff Bezos' Blue's Origin, to develop landers that can take astronauts to the lunar surface. Also, NASA has given a construction company $57 million contract to develop 3D printing technology to build roads, launchpads and homes on the lunar surface.

China

China started its Moon journey with the launch of Chang'e-1, an orbiter during October 2007. The first Chinese lunar landing happened during Chang'e-3 mission; it was a lander and rover system. Presently,

Chang'e-4 and Chang'e-5 systems are operational. China has succeeded with a sample return mission. The following table provides the relevant details of their various missions. They can reach the Moon quickly and do not travel in a circuitous fashion like India to reach the Moon.

Table 5
China's Missions to the Moon

Mission	Launch Date	Orbital Insertion	Landing Date	Remarks
Chang'e-1	24 October 2007	7 November 2007		orbiter mission
Chang'e-2	1 October 2010	6 October 2010		orbiter mission
Chang'e-3	1 December 2013	6 December 2013	14 December 2013	lander & rover
Chang'e-4	7 December 2018	12 December 2018	3 January 2019	lander & rover
Chang'e-5	23 November 2020	28 November 2020	1 December 2020	lander, sample return back

Source: https://tinyurl.com/bdcn8et7

Global Moon Agenda

China's Chang'e-6 (2024) is expected to perform a sample return from the South Pole Aitken basin on the far side of the Moon. Chang'e-7 (2026) is also expected to be a science mission to know more about the water on the lunar surface. During Chang'e-8 (2028) mission, they are keen to test technology for using in-situ resources and manufacturing with 3D printing. Finally, around 2030, China plans to undertake a human Moon mission.

Russia

Russia launched their mission to the Moon called Luna 25 on 11 August 2023 (Indian Standard Time). This was their first lunar lander mission in 47 years. It could even be viewed as the first Russian mission to the Moon, since all the earlier missions were undertaken during the Soviet era. Luna 25 was likely to reach the Moon in five days and remain in the lunar orbit for five to seven days before attempting a soft landing on the lunar surface. Since this craft was to land on the lunar south pole around the same time as Chandrayaan-3, there was some media hype that India and Russia were racing for the lunar south pole. Unfortunately, Luna 25 mission failed after successfully entering the lunar orbit. Possibly, the mission had moved in an unpredictable orbit owing to some communication issues and crashed on the lunar surface. The failure of this lander mission, which was planned to operate on the lunar surface for one year, is a loss to science.

Russia (as the erstwhile Soviet Union) is an old player of the Moon domain. The Luna Programme's first mission was in 1959. Subsequently, 24 other missions were carried out until 1976—including some failures. All these missions had different goals: flyby missions, missions involving the craft orbiting the Moon and some soft landings. Soviets have many achievements in the lunar field. They were the first to undertake a flyby mission to the moon, to have a lunar satellite and to have an impact (drop equipment to reach the lunar surface) on the moon. They were the first to take photographs of the far side of the moon. They also achieved the first lunar soft landing, deployed a lunar rover and undertook an analysis of lunar soil. In addition, they also brought back the first lunar sample to Earth.

Apart from the Luna Programme, the Soviets also had a programme called Zond, which focused on the Moon, Mars and Venus. This robotic spacecraft programme (1964–1970) had two spacecraft series, one for interplanetary exploration, and the other for lunar exploration. An interesting aspect of this programme was that during the Zond 5 mission, the first terrestrial organisms—two tortoises and some other lifeforms—survived their journey around the moon in a capsule and returned to Earth. Their human lunar programme (no achievements in this field) was kept secret and details of it were made public only in 1990. The Zond program had some focus on developing a human lunar programme. In addition, from 1969–1972, they undertook four tests (all unsuccessful) of the N1/L3 rocket, a super heavy-lift launch vehicle, possibly the Soviet counterpart to the US Saturn V, which was used for the manned Moon mission.

Russia is keen to carry forward the Soviet space legacy, which is probably why they are continuing with Luna programme. Today, despite the failure of Luna 25 mission, they are planning to continue with programmes for lunar exploration. The future Luna 26, Luna 27 and Luna 28 missions are expected to happen as planned and will serve as the next steps in their ambitious lunar programme. However, owing to Russia's invasion of Ukraine, the ESA has terminated cooperative activities with Russia on the impending Luna 26, Luna 27 and Luna 28 missions. ESA has also suspended the joint mission with Russia called the ExoMars. Luna 27 and Luna 28 are planned as sample collection and analysis, and sample collection and return missions respectively. ESA was to supply drilling and sampling tools to Russia. Now, it would be interesting to see, how Russia manages these missions without the ESA assistance.

Japan

Some decades back, Japan had articulated a major moon agenda. However, their progress has been slow. In 1960, their first lunar mission took place. At that time, the Hiten spacecraft had looped around the moon. SELENE (Selenological and Engineering Explorer), nicknamed Kaguya, was successfully launched during September 2007. Japan Aerospace Exploration Agency (JAXA) has articulated plans to have a rover and lander mission and a lunar sample return mission. During April 2023, Japan attempted the first private Moon landing, which ended in failure. This spacecraft called Hakuto-R was launched by SpaceX and took three months to enter the lunar orbit. However, the subsequent landing on the Moon's surface was a failure. It carried two lunar rovers: a four-wheeled one mini rover made by Sony and Japanese toy company Tomy, and one from the United Arab Emirates (UAE). Recently,

States like UAE and Saudi Arabia are taking major interests towards establishing space programmes. During the first week of September 2023, Japan launched its mission to the Moon. This is the first Moon landing attempt made by JAXA. This mission called SLIM (Smart Lander for Investigating Moon) is a small spacecraft, weighing just about 200 kg (the Chandrayaan-3 lander module weighed about 1,750 kg). This mission is expected to reach the moon by February 2024. The main objective of SLIM is to demonstrate precise landing, within 100 metres of the chosen site. This is going to be an attempt to validate pinpoint landing technology.

There are reports that ISRO is preparing for a lunar mission in collaboration with JAXA. This mission, called LUPEX, or Lunar Polar Exploration, is scheduled for 2024–25[38]. From the Indian front, there is no clarity regarding the future of India's Moon programme. Theoretically, the next mission, Chandrayaan-4, should be planned as a sample return mission.

South Korea

This technologically advanced State surprisingly took a good amount of time to become a spacefaring

State and made it to the coveted list only by June 2022. On 4 August 2022, South Korea launched Danuri or the Korean Pathfinder Lunar Orbiter (KPLO) to the Moon. This satellite was inserted into orbit around the Moon on 16 December 2022. There was support from NASA for undertaking this mission and the orbiter was launched by SpaceX. South Korea is sending a lunar monitoring payload to the US for carrying out science experiments as part of the still unnamed US moon exploration mission likely to happen during 2024. This Lunar Space Environment Monitor (LUSEM) is a sensor for detecting high-energy particles on the lunar surface and has been developed by the Korea Astronomy and Space Science Institute.

North Korea

As per some reports, Democratic People's Republic of Korea (DPRK) is also having a programme called North Korean Lunar Exploration Programme (NKLEP). They want to have their own lunar orbiter and want to go for a lunar lander and subsequently undertake a lunar sample return mission. Nothing much is known about the exact status of their programme.

Israel

The State has its own space programme with limited mandate. In the past, there were arguments regarding Israel not being interested in undertaking glamorous programmes like Moon missions. However, recently, Israel supported a private agency to establish a Moon programme. Beresheet was Israel's first lunar mission and the first attempt by SpaceIL, a private company, to land on the Moon. The launch was successful, and the mission did achieve lunar orbit, but was lost during an April 2019 lunar landing attempt. This agency is planning for its second mission around 2024.

There can be no debate that the return of humans to the moon is going to become a reality in the upcoming years. Today, various Moon programmes in the world are at various levels of development. Private agencies are yet to get success with their moon agendas. It appears that the moon would be eventually conquered by using a public-private partnership model.

8

Moon Rovers: An Overview[39]

People in general remember the achievers, who accomplish something first. Hence, Apollo 11 is the most discussed Moon mission till date. The last three Apollo missions (15, 16 and 17), which happened during 1971 and 1972, were also important from a scientific perspective. These missions gave humans better exposure to what this natural satellite of Earth is made of.

Apollo Missions

Apollo 11 was all about geopolitics. However, subsequent human missions to the Moon did undertake some experiments and more than that, a major focus

Moon Rovers: An Overview

was on lunar sample collection. During these three missions, the US astronauts travelled on the Moon's surface by sitting in a Moon buggy (also known as Lunar Roving Vehicle). Chandrayaan-3's rover is known to have moved on the lunar surface and covered 100 m during its movement from 23 August to around 2 September 2023. Pragyan is a robotic equipment, while the Moon buggy, which moved on the lunar surface during Apollo (15, 16 and 17) missions could be viewed as a human operated rover.

Before the success of Chandrayaan-3, there were only three countries who succeeded in operating rovers on the Moon's surface: the US, Russia (erstwhile Soviet Union) and China. For the US, their astronauts have driven the rovers on the Moon's surface, while in the case of the Soviets and China, the rovers functioned as robotic equipment. Among the three, it was the Soviet Union which became the first State in making rovers travel on the Moon's surface. They undertook two rover missions during the 1970s.

Lunokhod Missions

For the Soviets, their Luna 17 spacecraft (1970) carried the Lunokhod 1 rover to the Moon. This was the first wheeled craft ever operated on another celestial body. The earlier 1969 attempt, Lunokhod 0), was a failure. Lunokhod 1 operated on the lunar surface for 11 lunar days. This mission was designed to last for around 90 Earth days, but the rover remained operational for 321 Earth days. During this stay, it travelled a total distance of 10.54 km. Apart from antennas and batteries, Lunokhod 1 was equipped with four television cameras, devices to test the lunar soil, an X-ray spectrometer, X-ray telescope, cosmic ray detectors and a laser retroreflector (supplied by France). In April 2010, the present location of Lunokhod 1 was found by scientists from the US. They used the laser ranging techniques for this purpose. In 2010, the US scientists and in 2013, the French scientists are known to have used Lunokhod 1's reflector for some experiments indicating that after a silence of more than four decades, this rover still had a lot to offer.

Lunokhod 2 was the second Soviet robotic lunar rover, which was flown by the Luna 21 spacecraft in 1973. This 1,814 kg rover is known to have survived for about four months and travelled a distance between 37 to 42 km on the lunar surface. Some experiments to understand the local magnetic fields and the soil mechanics of the lunar surface material were carried out. There is an interesting story about this mission.

Moon Rovers: An Overview

During December 1993, at a Sotheby's auction, Lunokhod 2 and the Luna 21 lander were sold for $68,500. The buyer Richard Garriott was a computer game magnate. He is America's first second-generation astronaut. His father was a NASA astronaut, and he paid money to stay at the International Space Station (ISS) for two weeks in 2008. For all these years, the US space agency NASA has successfully undertaken some amazing missions. Today, NASA has an operational helicopter flying on Mars called Ingenuity (or Ginny), and on 19 April 2021 it successfully undertook the first powered controlled extraterrestrial flight by an aircraft. The copter has been operational for more than 865 days and has undertaken 58 flights as of 11 September 2023. Another amazing mission by NASA involves the Lunar Roving Vehicle (LRV) moving on the Moon's surface.

References to the lunar rover are found in scientific fiction published during the 1950s. NASA is known to have been working since the 1960s on this idea. Before the launch of Apollo 11, the private aerospace company Boeing started conceptualizing the idea of the LRV. Subsequently, NASA successfully operated a battery-powered four-wheeled rover (developed by Boeing) on the Moon during the Apollo (15, 16 and 17) missions. Before the Apollo 15 mission, the astronauts used to walk in the area close to the landing site of the spacecraft, however with the availability of the Moon buggy, they were able to visit a longer distance and collect more information and rock samples. At the same time, it was ensured that, at a time, the astronauts covered only the distance which they are capable of walking back. This was decided so that the astronauts would not face any problems in case their rover vehicle breaks down. The total distances covered during Apollo 15 and Apollo 16 missions were around 27 km during each mission, while Apollo 17 covered a total of 36 km. It is important to note that since humans were involved in these missions, they were able to make decisions about the travel locations and the rocks to be picked up to get them back on Earth. These vehicles used to weigh 210 kg and were designed to hold an additional payload of 230 kg. They are left abandoned on the Moon's surface.

Chang'e Missions

China did the first soft landing on the lunar surface after the 1973 Lunokhod 2 mission. Their Chang'e-3 mission (2013) to the Moon carried the 140 kg rover called Yutu. The rover's primary task was to undertake

Moon Rovers: An Overview

lunar surface topography and geological survey. The payloads on the rover included a ground-penetrating radar and spectrometers. With this, for the first time in history, knowledge about the lunar structure to a depth of 30 m and the lunar crust structure down to more than 100 m deep was gained. This rover made the first lunar surface contact on 14 December 2013, and the rover remained operational till December 2015.

On 3 January 2019, China's Chang'e-4 successfully landed on the far side of the moon. The mass of the lander is about 1,200 kg and that of the rover (Yutu-2) is 140 kg. The rover has been exploring the lunar surface now for more than four years and is in good health and regularly transmitting data. The information received from this rover indicates that there is much difference in ground conditions on the far side of the Moon than the other parts. Here, the soil is sticky and more conducive to conducting activities. The area has an abundance of small rocks and impact craters.

It is surprising to note how the activity on the moon was far better some five decades back as compared to today. To undertake research on other planets, robotic activity is the most preferred option and rightfully so. This is because it is dangerous to risk the human life when there are many unknowns about the place where you are visiting. Also, the cost of a mission increases significantly when humans are involved. On the other hand, look at the advantages: during 1970s NASA astronauts had covered more than 35 km and collected many rock samples. In comparison, Chandrayaan-3's rover could cover only 100 m in around 10 to 14 days of time. China's Yutu-2 rover released by Chang'e-4 during 2019 is possibly still operational, however, even during such a long duration, it has travelled only a distance between 1.5 to 2 km. All this indicates that if we want to know more about the Moon in shortest possible time, then having a human mission is a better option.

9

India's Moon Tryst[40]

Chandrayaan-3's success reflects the grit and determination of ISRO's scientific community. Post 2019, they worked systematically towards ensuring that nothing would be left to chance when they reattempted the soft landing on the lunar surface.

For India, this is not only a demonstrative mission to showcase their expertise regarding soft landing on the Moon but also has some scientific agenda associated with it. Already some initial reports have been received from the lander and rover providing details about the thermal profile of the Moon surface and beneath, types of minerals available on the lunar surface and about plasma densities above the lunar surface.

Chandrayaan-3

India's Soft Power

The success of this mission is also good for India's commercial space sector, and it demonstrates the capability of the Indian system to pull off such technology intensive projects. All this is expected to attract more investments to the space sector. From an economic perspective, space is emerging as a promising sector for the future and it could be said that India's space programme has blossomed timely. India's Chandrayaan-3 is important from a scientific, technological and commercial point of view. India comes from a grouping of developing countries which are known to be part of the Global South. The geopolitical implications of this success could be viewed as India getting a 'soft power' prestige.

Is India a part of the so-called 'moon race'? There is no reason for India to get into this 'race'. Just because some States are trying to reach the Moon roughly around the same period, it should not be perceived as India trying to win a race. It is important to realize that the competition in space was a defining part of the Cold War era's power politics. Today, every State is planning their space programmes based on their technological and financial capabilities. It is well understood that States are going to the Moon for planetary resources. But at the same time, States understand that going solo is not a practical option and that's why ideas like the Artemis Program are put in place. Even China and Russia are keen to establish something like a Moon corridor. Today, Moon-faring States have different levels of expertise; some are just beginners while some have much-developed programmes. To say that a State like India aspiring to put a spacecraft into lunar orbit is in a race with another that has already undertaken a successful Moon sample return mission is deceptive.

What Lies Ahead

Given present trends, mostly it would be the US and China who would be fighting it out in the planetary ring. States like India should keep themselves away from any such rat race. The success of Chandrayaan-3 tells us about the capabilities of ISRO and what they can achieve in future. Also, ISRO has good relations with all major spacefaring States and should think towards planning joint collaborations. Now, time has come for India to clearly enunciate its Moon agenda and plan her future course of action.

Notes

Chapter 1: India Scripts History

1. Welle, Deutsche, 'Osiris-REx and Bennu: Sample return missions explained', *Frontline*, 25 September 2023, https://frontline.thehindu.com/news/osiris-rex-and-bennu-sample-return-nasa-asteroid-missions-explainer/article67344853.ece. Accessed on 26 September 2023.
2. Pal, Sanchari, 'Transported on a Bicycle, Launched from a Church: The Amazing Story of India's First Rocket Launch', *The Better India*, 8 November 2016, https://www.thebetterindia.com/74283/first-rocket-india-thumba-vikram-sarabhai-abdul-kalam/. Accessed on 22 September 2023.
3. 'Dr. Vikram Ambalal Sarabhai (1963-1971)', *Indian Space Research Organisation (ISRO)*, https://www.isro.gov.in/sarabhaiformer.html. Accessed on 22 September 2023.

Notes

4. Mehta, Jatan, 'Chandrayaan-1, India's first Moon mission', *The Planetary Society*, https://www.planetary.org/space-missions/chandrayaan-1. Accessed on 15 September 2023.
5. Ibid.
6. Datta, Jayati, and S.C. Chakravarty, *Chandrayaan-1: India's First Mission to Moon*, Indian Space Research Organisation, Bangalore.

Chapter 2: The 'Dynamics' of the Moon

7. Theia (about the size of Mars) is a hypothesized ancient planet in the early Solar System. As per the giant-impact hypothesis, it collided with the early Earth around 4.6 billion years ago. This collision led to debris gathering which eventually led to the formation of the Moon.
8. In 2018, Sotheby's auctioned the only government-certified sample of loose moon dust in private hands, three tiny pebbles brought back by the former Soviet Union's Luna 16 robotic probe in 1970. The grains, which weighed a total of about 0.2 grams, were sold for $855,000.
9. The dark areas found on the lunar surface are known as 'mare' (or maria) which is the Latin word for 'seas'. Centuries ago, scientists mainly based on telescopic observations believed that there could be oceans on the moon. But the visible dark stretches on the Moon's

surface were not oceans. Today we know that the 'maria' are the volcanic basins

10. For this chapter, these sources have been referred to for facts on the history of the Moon, its phases and geology:
    Buckle, Anne, and Graham Jones, 'The Phases of the Moon', *timeanddate*, https://www.timeanddate.com/astronomy/moon/phases.html. Accessed on 22 September 2023
    'When Was the Moon Discovered?', *The Nine Planets*, 22 August 2023, https://nineplanets.org/questions/when-was-the-moon-discovered/. Accessed on 22 September 2023.
    'Geology Of Moon', *Geology In*, https://www.geologyin.com/2014/12/geology-of-moon.html. Accessed on 22 September 2023.
    'In Depth | Earth's Moon—NASA Solar System Exploration' *NASA*, https://solarsystem.nasa.gov/moons/earths-moon/in-depth/. Accessed on 22 September 2023.

11. Noble, Sarah, 'The Lunar Regolith', *NASA*, https://www.nasa.gov/sites/default/files/atoms/files/05_1_snoble_thelunarregolith.pdf. Accessed on 9 September 2023.

12. The entire discussion is based on:
    Haxton, Tom. 'What Resources Could We Find on the Moon? Here are Three Possibilities!' *HeroX*, https://www.herox.com/blog/954-what-resources-could-we-find-on-

Notes

the-moonhere-are. Accessed on 22 September 2023.
Staedter, Tracy, 'Why on Earth Should We Be Mining the Moon?', *Northrop Grumman*, 3 September 2020, https://now.northropgrumman.com/why-onearth-should-we-be-mining-the-moon. Accessed on 22 September 2023.
'Explainer: Why are space agencies racing to the moon's south pole?', *Reuters*, 24 August 2023, https://www.reuters.com/science/why-are-space-agencies-racing-moons-southpole-2023-08-22/. Accessed on 22 September 2023.
Patel, Neel V., 'Here's how we could mine the moon for rocket fuel', 19 May 2020, *MIT Technology Review*, 19 May 2020, https://www.technologyreview.com/2020/05/19/1001857/how-moon-lunar-mining-water-ice-rocket-fuel. Accessed on 22 September 2023.

Chapter 3: India's First Two Missions to the Moon

13. Drake, Nadia, 'Where, exactly, is the edge of space?', *National Geographic*, 21 December 2018, https://www.nationalgeographic.co.uk/space/2018/12/where-exactly-is-the-edge-of-space. Accessed on 25 September 2023.
14. Howell, Elizabeth, 'What is Space', *Space.com*, 17 February 2022, https://www.space.com/24870-what-is-space.html. Accessed on 21 September 2023.

15. R, Mike, 'Deep Space: What Exactly Does It Mean?', *Cosmonpnw*, 6 January 2021, https://cosmospnw.com/deep-space-what-exactly-does-it-mean/. Accessed on 21 September 2023.
16. Datta, Jayati, and S.C. Chakravarty, *Chandrayaan-1: India's First Mission to Moon*, Indian Space Research Organisation (ISRO), Bangalore.
17. Howell, Elizabeth, 'Chandrayaan-1: India's First Mission to the Moon', *Space.com*, 28 March 2018, https://www.space.com/24870-what-is-space.html. Accessed on 21 September 2023.
18. States like the US, Japan and China are in the hunt for deposits of Helium-3, which could help to produce waste-free nuclear energy that could be answer many of Earth's energy problems
19. Meheta, Jatan, 'Chandrayaan-1, India's first Moon mission', *The Planetary Society*, https://www.planetary.org/space-missions/chandrayaan-1,. Accessed on 21 September 2023.
20. 'Chandrayaan-1_Science', *Indian Space Research Organisation (ISRO)*,
https://www.isro.gov.in/Chandrayaan-1_science.html#:~:text=Salient%20science%20results%20from%20 Chandrayaan%2D1&text=Three%20dimensional%20conceptualization%20of%20many, Moon%20(TMC%20and%20HySI). Accessed on 21 September 2023.

Notes

21. 'Science Results from Chandrayaan 2 Mission', *Indian Space Research Organisation (ISRO)*, August 2021, https://www.isro.gov.in/media_isro/pdf/ResourcesPdf/science_results_from_ch-2.pdf. Accessed on 21 September 2023.
22. 'Chandrayaan-2 (isro.gov.in)', *Indian Space Research Organisation (ISRO)*, https://www.isro.gov.in/Chandrayaan2_science.html. Accessed on 21 September 2023.

Chapter 4: Chandrayaan-3: ISRO's Moon Supremacy

23. 'Chandrayaan-3 Details', *Indian Space Research Organisation (ISRO)*, https://www.isro.gov.in/Chandrayaan3_Details.html. Accessed on 21 September 2023.
24. The various data about the mission mentioned in the section 'Chandrayaan-3' mission profile is available at the official website of ISRO:
'Chandrayaan-3(isro.gov.in)', *Indian Space Research Organisation (ISRO)*,
https://www.isro.gov.in/Chandrayaan3.html#:~:text=Lunar%2DOrbit%20Insertion%20(LOI),planned%20for%20Aug%205%2C%202023.&text=Orbit%2Draising%20maneuver%20performed%20on,planned%20for%20August%20

1%2C%202023.&text=The%20fourth%20 orbit%2Draising%20maneuver,bound%20perigee%20 firing)%20is%20completed. Accessed on 21 September 2023.

25. C S, Hemanth, 'Chandrayaan-3 | How Nasa, ESA, will support ISRO during the Moon landing on August 23', *The Hindu*, 20 August 2023, https://www.thehindu.com/sci-tech/science/chandrayaan-3-how-nasa-esa-will-support-isro-during-the-moon-landing-on-august-23/article67216311.ece. Accessed on 21 September 2023.

26. 'Chandrayaan-3 Details', *Indian Space Research Organisation (ISRO)*, https://www.isro.gov.in/Chandrayaan3_Details.html. Accessed on 21 September 2023.

Chapter 5: Learning from Chandrayaan-2's failure

27. Meheta, Jatan, 'Explainer: Why Did the Chandrayaan-2 Lander Take so Long to Find?', *The Wire Science*, 14 December 2019, https://science.thewire.in/aerospace/explainer-why-did-the-chandrayaan-2-lander-take-so-long-to-find/. Accessed on 21 September 2023.

28. Lele, Ajey, 'Chandrayaan 2's Moon illusion', *The Space Review*, 16 September 2019, https://www.thespacereview.com/article/3793/1. Accessed on 21 September 2023.

29. Ibid.

Notes

30. Dixit, Pranav, "15 minutes of terror': Chandrayaan-3's daring lunar touchdown effort explained', *Business Today*, 23 August 2023, https://www.businesstoday.in/technology/news/story/15-minutes-of-terror-chandrayaans-daring-lunar-touchdown-effort-explained-395176-2023-08-23. Accessed on 21 September 2023.

31. Anand, Nisha, 'Chandrayaan-3: Why did Chandrayaan-2 lander fail? Why is it so difficult to land on Moon', *Hindustan Times*, 14 July 2023, https://www.hindustantimes.com/india-news/chandrayaan3-mission-latest-news-moon-landing-challenges-isro-chandrayaan-3-launch-live-sriharikota-location-101689314479570.html. Accessed on 21 September 2023.

32. Singh, Surendra, 'Why did Chandrayaan-2 lander fail? ISRO chief gives 3 key reasons for crashlanding', *Times of India*, 12 July 2023, https://timesofindia.indiatimes.com/india/why-did-chandrayaan-2-lander-fail-isro-chief-gives-3-key-reasons-for-crashlanding/articleshow/101678204.cms?from=mdr&utm_source=contentofinterest&utm_medium=text&utm_campaign=cppst,. Accessed on 21 September 2023.

Chapter 6: Industry and People

33. 'Aerospace Division of HAL-Brief Profile and Core Capabilities', *Indian Space Research Organisation (ISRO)*,

https://www.isro.gov.in/g20selm/assets/pdf/OptionA_Hindustran_Aeronautics_Limited_Aerospace_Division.pdf. Accessed on 21 September 2023.

34. 'MIDHANI's Contribution in Chandrayaan-3', *Mishra Dhatu Nigam Limited*, https://midhani-india.in/news_marquee/midhanis-contribution-in-chandrayaan-3/. Accessed on 21 September 2023.

35. 'MSME ministry contributed significantly in realising Chandrayaan-3 dream: Rane', *Press Trust of India*, 23 August 2023, https://www.ptinews.com/news/business/msme-ministry-contributed-significantly-in-realising-chandrayaan-3-dream-rane/636100.html. Accessed on 21 September 2023.

36. For the data on scientists involved with Chandrayaan-3, these sources have been referred to:

    Saumya, 'List of All Scientists Behind the Historic Chandrayaan-3 Mission Success', *Jagran Josh*, 28 August 2023, https://www.jagranjosh.com/general-knowledge/scientists-behind-chandrayaan-3-success-1692796298-1, Accessed on 21 September 2023.

    'Chandrayaan-3: Meet the architects behind India's Moon mission', *The Economic Times*, 24 August 2023, https://economictimes.indiatimes.com/news/science/chandrayaan-3-meet-the-architects-behind-indias-moon-mission/articleshow/102972258.cms?utm_ source=contentofinterest&utm_medium=text&utm_campaign=cppst. Accessed on 11 September 2023.

Chapter 7: Global Moon Agenda

37. Some of the information in this chapter is from author's earlier publications (Various websites have been referred to for specific data on various missions discussed in the chapter):
'Global Moon Agenda', *Financial Express*, 9 July 2023. https://www.financialexpress.com/business/defence-global-moon-agenda-3161848/. Accessed on 21 September 2023.
Lele, Ajey, 'What Russia's Lunar Mission Tells Us About Its Future Plans', *The Wire*, 14 August 2023, https://thewire.in/space/russia-moon-agenda-luna-25. Accessed on 21 September 2023.
Lele, Ajey, 'Will Russian Luna 25 beat Chandrayaan-3 to land first on the Soth Pole of the Moon', *taazakhabar news* https://taazakhabarnews.com/will-russian-luna-25-beat-chandrayaan-3-to-land-first-on-the-south-pole-of-the-moon/. Accessed on 11 September 2023.
38. Sengupta, Arjun, 'Sun mission, Japan collaboration and more: What has ISRO planned after Chandrayaan-3's success?', *The Indian Express*, 27 August 2023, https://indianexpress.com/article/explained/explained-sci-tech/isro-future-missions-8910764/. Accessed on 21 September 2023.

Chapter 8: Moon Rovers: An Overview

This chapter is an updated version of the author's article cited below.

39. Lele, Ajey, 'Moon Rovers-impressive past and the present', *taazakhabar news*, https://taazakhabarnews.com/moon-rovers-impressive-past-and-present/. Accessed on 21 September 2023

Chapter 9: India's Moon Tryst

Some of the facts and information in this chapter are taken from the author's article cited below:

40. Lele, Ajey, 'India is on the Moon but needs to avoid the "Moon Race" trap', *The Space Review*, 5 September 2023, https://www.thespacereview.com/article/4645/1 Accessed on 21 September 2023

Index

Alpha Particle X-ray Spectrometer (APXS), 70, 72, 73
antenna, 69
Apollo, 18, 20, 21, 24, 26, 97, 99, 110, 111, 113, 114
Apollo 11, 21, 24, 110, 113
Apollo 12, 24
Apollo 13, 99
Apollo 14, 24
Apollo 15, 24, 113, 114
Apollo 16, 24, 114
Apollo 17, 24, 114
Artemis 1, 99
Artemis 2, 100, 101
Artemis 3, 101
Artemis Accords, 98
Artemis Program, 98, 118
artificial intelligence (AI), 82, 83
Astronautical Society of India (ASI), 9

Bennu, 2
Beresheet, 109
Boeing, 32, 113
Breccias, 22
Buzz, 24

C1XS, 43
Callisto, 2
Capture theory, 17
ChACE, 46, 51
ChACE-2, 51
Chandra's Surface Thermophysical Experiment (ChaSTE), 70, 71, 73, 74, 89
Chandrayaan-1, 11, 40, 41, 42, 43, 45, 46, 49, 58
Chandrayaan-2, 40, 48, 49, 50, 51, 52, 53, 57, 58, 62, 66, 68, 76, 77, 78, 83, 84, 85, 95
Chandrayaan-2 Large Area Soft X-ray Spectrometer

(CLASS), 50
Chandrayaan-3, vii, viii, 52, 53, 56, 57, 58, 59, 61, 62, 64, 65, 66, 69, 70, 73, 74, 78, 84, 85, 86, 88, 89, 90, 91, 92, 93, 94, 95, 96, 99, 103, 107, 111, 116, 117, 118, 119
Chang'e-1, 101, 102
Chang'e-3, 101, 102, 114
Chang'e-4, 102, 115, 116
Chang'e-5, 20, 102
Chang'e-6, 103
Chang'e-7, 103
Chang'e-8, 103
China, vii, 14, 20, 27, 32, 37, 98, 101, 102, 103, 111, 114, 115, 116, 119
Clementine, 46
Coaccretion theory, 17
Cold War, 6, 25, 118
Compton Belkovich Volcanic Complex, 47
core, 18, 20, 22, 61, 93
crust, 20, 21, 22, 41, 71, 115

deep space, 1, 9, 33, 38, 39, 40, 68, 69, 83, 84, 95
deorbiting manoeuvres, 79
DFSAR, 51
DSN, 69
DSS-34, 69
DSS-36, 69
DSS-65, 69

Earth days, 12, 14, 52, 59, 112
Earth orbit manoeuvres, 62
Edwin 'Buzz' Aldrin, 24
Elliptic Parking Orbit (EPO), 64
energy crisis, 32
ESA, 2, 27, 43, 44, 50, 69, 98, 105, 106
Europa, 2
ExoMars, 105

Fission theory, 17
flyby, 20, 27, 98, 100, 101, 104

Gaganyaan, 61, 62
Galileo, 17
Ganymede, 2
Garriott, Richard, 113
geology of the Moon, 20
giant-impact hypothesis, 18
gravity, 20, 34, 83, 84
GSLV, 7, 61

Hakuto-R, 106
hard landing, 50, 76
Helium-3, 31, 32
High Energy X-ray Spectrometer (HEX), 43
Hindustan Aeronautics Limited (HAL), 88, 89
Hiten spacecraft, 106

Index

Hyper Spectral Imager (HySI), 43, 47
IDSN, 45, 70
Imaging IR Spectrometer (IIRS), 51
Indian National Committee for Space Research (INCOSPAR), 4
Indian National Space Promotion and Authorisation Centre (IN-SPACe), 87
Indian Space Association (ISpA), 88
Ingenuity (or Ginny), 113
in-situ, 30, 31, 32, 34, 59, 103
Instrument for Lunar Seismic Activity (ILSA), 70, 71
International Space Station (ISS), 34, 113
Israel, 27, 98, 109
ISRO, vii, 4, 6, 7, 8, 9, 10, 11, 40, 42, 43, 44, 45, 46, 49, 50, 52, 53, 54, 56, 57, 58, 61, 62, 63, 66, 67, 68, 69, 70, 73, 74, 75, 76, 77, 78, 81, 84, 85, 86, 88, 89, 94, 95, 96, 100, 107, 117, 119
ISTRAC, 45, 64, 95

James Webb Space Telescope, 2
Japan, 3, 27, 98, 106, 107
Japanese agency Institute of Space and Astronautical Science (ISAS), 27
JAXA, 106, 107
Jupiter Icy Moons Explorer (Juice), 2

Kármán line, 37
Kasturirangan Dr Krishnaswamy, 8
Kepler, Johannes, 16
K. Kalpana, 95
Korean Pathfinder Lunar Orbiter (KPLO), 108
KREEP, 54
Kumar, S. Mohana, 95

lander, 5, 12, 49, 50, 52, 57, 59, 60, 63, 66, 67, 68, 70, 73, 74, 76, 77, 78, 79, 81, 82, 84, 85, 86, 89, 92, 101, 102, 103, 104, 106, 107, 108, 113, 115, 117
lander module (LM), 59, 65
lander special tests, 60
Langmuir Probe, 70
Laser Induced Breakdown Spectroscope (LIBS), 70, 72, 73
Laser Inertial Referencing & Accelerometer Package (LIRAP), 92
Laser Retroreflector Array, 70

launch vehicle, 59, 61, 87, 88, 90, 91, 93, 95, 105
Luna 1, 24
Luna 17, 111
Luna 21, 112, 113
Luna 25, 103, 105
Luna 26, 105
Luna 27, 105
Luna 28, 105
Luna programme, 20, 27, 105
lunar day, 12, 52, 59, 73
Lunar dust, 23
lunar exploration, 97, 104, 105
lunar geologic, 21
Lunar Laser Ranging Instrument (LLRI), 43, 47
Lunar Orbit Insertion (LOI), 65, 79, 100
lunar orbit manoeuvres, 63
lunar orbit reduction manoeuvres, 42
lunar origin, 17, 18
Lunar Reconnaissance Orbiter (LRO), 47, 77
Lunar Roving Vehicle (LRV), 113
lunar soil, 23, 49, 104, 112
lunar surface, vii, 12, 19, 20, 21, 22, 24, 31, 34, 41, 45, 46, 47, 48, 50, 51, 53, 57, 59, 64, 65, 66, 68, 71, 73, 74, 75, 76, 77, 83, 84, 99, 101, 103, 104, 108, 111, 112, 114, 115, 117
Lunokhod 0, 112
Lunokhod 1, 111, 112
Lunokhod 2, 112, 113, 114
LUPEX, 107
LVM-3, 61, 62, 64, 91
LVM3, 59, 61, 89, 91, 93

M3, 44, 46, 47
magma ocean, 21, 41
mantle, 18, 19, 20, 22, 71
maria, 22
Mars, 30, 39, 40, 45, 95, 96, 104, 113
megaregolith, 23
microflares, 53, 55
mineral load, 34
Mini-SAR, 44, 45, 47
mission objectives, 48, 59, 60
Moon buggy, 111, 114
Moon corridor, 119
Moon Impact Probe (MIP), 42, 43, 46, 47
moon mining, 30
Moon phases, 15
moon race, 30, 118

Nair, S. Unnikrishnan, 95
NASA, 2, 20, 24, 26, 27, 39, 44, 45, 46, 47, 50, 58, 69, 70, 77, 98, 99, 101, 108, 113, 116
National Aerospace Laboratories, 88

Index

National Lunar Mission Task Force, 10
Neil Armstrong, 7, 24, 26
NewSpace India Limited, 87
NKLEP, 108
North Atlantic Treaty Organization (NATO), 25
North Korea, 98, 108

orbit, 1, 7, 8, 14, 15, 16, 17, 26, 39, 40, 42, 49, 53, 55, 59, 61, 62, 63, 64, 65, 78, 79, 80, 95, 99, 100, 103, 104, 106, 108, 109, 119
orbit burn, 42
orbiter, 8, 11, 27, 40, 42, 46, 49, 50, 52, 53, 66, 68, 77, 78, 81, 86, 98, 101, 102, 108
Orbiter High Resolution Camera (OHRC), 51
orbit-raising manoeuvres, 49, 62, 78
Orion, 99, 100
OSIRIS-REx, 2
outer space, 25, 36, 37

payload, 50, 59, 108, 114
period of terror, 65
PM, 59
Pragyan rover, 50, 74
private space industry, 87, 88
probe, 39, 71, 72, 73, 74
propulsion module (PM), 59, 63, 65

proto-Moon, 17, 18, 19
PSLV, 7, 8, 9, 42

Radio Anatomy of Moon Bound Hypersensitive ionosphere and Atmosphere (RAMBHA), 51, 71, 89
RADOM-7, 44
Ramakrishna, B.N., 95
RAMBHA-DFRS, 51
Rare Earth Elements (RREs), 32
Rare Earth Metals (REMs), 32
regolith, 22, 23, 53
Roscosmos, 48
rover, 5, 12, 49, 50, 52, 57, 59, 68, 70, 73, 74, 76, 78, 92, 101, 102, 104, 106, 111, 112, 113, 114, 115, 116, 117
Russia, vii, 3, 25, 48, 49, 62, 98, 103, 104, 105, 106, 111, 119

sample return mission, 20, 74, 102, 106, 107, 108, 119
Sankaran, M, 95
SARA, 43, 47
SatCom Industry Association (SIA-India), 88
satellite, 6, 7, 8, 9, 13, 37, 39, 40, 61, 69, 78, 88, 89, 91, 104, 108, 110
Satish Dhawan, 6, 59

Saudi Arabia, 27, 107
SELENE (Selenological and Engineering Explorer), 106
selenophile, 56
Shanmuga Subramanian, 77
SIR-2, 44
Sivan, K., 81
SLIM, 107
SLS rocket, 99
soft landing, vii, 49, 50, 52, 59, 63, 66, 103, 104, 114, 117
Somanath, S., 78, 94
sounding rocket, 3, 5
South Korea, 27, 98, 107, 108
south pole, vii, 33, 43, 46, 52, 53, 67, 73, 103
South Pole Aitken (SPA), 54
Soviet Union, vii, 3, 7, 20, 23, 104, 111
space programme, vii, 3, 4, 5, 9, 109, 118
SpaceX, 101, 106, 108
spectrometer, 43, 44, 53, 55, 112
Spectro-polarimetry of Habitable Planet Earth (SHAPE), 72
Srivastava, Ritu Karidhal, 96
SSLV, 7
stratigraphy, 41

telescope, 2, 35, 38, 112
Terrain Mapping Camera-2 (TMC-2), 51
Terrain Mapping Camera (TMC), 43, 47, 51, 54
The US, 98
Thumba, 3, 4, 5
translunar injection, 49, 63, 64, 99, 100
Tycho crater central peak, 47

UAE, 27, 106, 107
V
Vanitha, Muthayya, 95
Veeramuthuvel, P, 95
Vikram lander, 50, 73, 77
Vikram Sarabhai, 3, 5, 95
Vikram Sarabhai Space Centre, 5, 95

water ice, 33, 46, 48, 51, 52, 67

XSM, 51, 54, 55

Yuri Gagarin, 7, 25
Yutu, 114, 115, 116
Yutu-2, 115, 116

Zond, 104, 105

www.ingramcontent.com/pod-product-compliance
Lightning Source LLC
Chambersburg PA
CBHW031416210526
45464CB00005B/1919